# A Gloucestershire Notebook

# A Gloucestershire
# Notebook

## Humphrey Phelps

To Hathrael again
For the same good reasons

First published in this form by The History Press in 2008

The Mill, Brimscombe Port, Stroud, Gloucestershire, GL5 2QG
www.thehistorypress.co.uk

Copyright © Humphrey Phelps, 2008

The right of Humphrey Phelps to be identified as the Author of this Work has been
asserted in accordance with the Copyrights, Designs & Patents Act 1988.

British Library Cataloguing in Publication Data:
A catalogue record for this book is available from the British Library

ISBN 978 1 84588 629 5

Typesetting and origination by The History Press
Printed and bound in Great Britain by
Ashford Colour Press Ltd, Gosport, Hampshire

# Preface

*A Gloucestershire Notebook* is a collection of articles written by Humphrey Phelps between 1976 and 1986. They were first published in the magazine *Gloucestershire & Avon Life*.

# Hard winter work, and harder hands

January 1986

When I left school 42 years ago, threshing corn was one of the winter jobs on the farm. The County War Agricultural Executive Committee, which everyone called the War Ag, compelled all farms of more than thirty acres to grow a certain amount of corn, and potatoes, sugar beet and milk production were also encouraged, in preference to meat and eggs. A farmer with no dairy herd was under particular pressure to grow a much larger percentage of arable crops.

In our predominantly livestock district of West Gloucestershire, nature and tradition led to a general hatred of potatoes and sugar beet. It was all hard work, then, with no machines to harvest these crops, apart from a bouting plough. This lifted the beet a little, but it was mainly a job for the muscle of man—and yes, the muscle of the land-girls.

Then there were the days and days spent topping and cleaning the beet with frozen hands, and with feet ankle-deep in mud. It must have been great hardship for the city land-girls who had been used to working in offices and shops. Sometimes, too, we would be helped by Italian prisoners of war, who could still forget about being far from their sunny home and captive in a foreign land, and burst into song in our bleak, misty fields.

Threshing was not a pleasant task, either; it was dirty and dusty, the workers sweated or froze in the grime, according to the job they were doing, and all the while the drum droned on remorselessly, like some huge monster with an insatiable appetite. Occasionally its guts rumbled and growled, when an uncut sheaf was dropped by mistake into its ever-open mouth. At one end it poured forth corn; at the other, threshed straw; and from beneath it disgorged cavings, all to a continuous belch of dust.

The day seemed interminable. As the rick of corn shrunk lower, the new rick of straw towered higher; the bags of corn were tied and wheeled away on a sack truck, and the cavings borne off on a split-open sack. Attending to the cavings was the dirtiest and lowliest of all the tasks of threshing.

Nine people were needed for threshing, including the driver and his mate who came with the travelling machine, the mate's job being to feed the sheaves into the drum. Two men were needed on the corn rick; another cut the bonds of the sheaves and handed them to the mate, and as both stood on top of the thresher, they were subjected to the cold winter wind, which could be more cruel inside a Dutch barn than at an outside stack.

Two more men attended to the threshed straw which fell from the drum into a machine which tied it into boltens; another looked after the corn which poured from the drum into four-bushel sacks, stoutly-made sacks hired from Gopsill Brown's of Gloucester Docks, and thus always known as Gopsills. Finally, there was the unfortunate who dealt with the cavings.

As the corn stack diminished, rats would appear. By law, wire netting had to be put up to prevent their escape, and while in this respect it was ineffective, it succeeded well as an encumbrance to the threshing gang. By law, too, in those wartime years, the driver had to keep a count of the sacks of corn. Oats, barley, beans and peas could be kept on the farm for feeding stock, but wheat had to go for human food; woe betide any farmer discovered with an undeclared sack of wheat on his premises.

Time, for me, has not cast a rosy glow on those days spent threshing, yet there was a certain jollity about it in spite of the unpleasantness, a jollity caused by the companionship and the cider. We drank a lot of cider; it helped dull the senses and make the dust bearable, and every so often the driver would bring the jar to each of us and pour out a draught into an oily, dusty mug, carefully removing any floating caving or thistledown with a grimy finger. Rough, sharp cider it was, and it set your teeth on edge, but to us it was ambrosia.

When I come to think of it, it was threshing that was really the cause of my leaving school a couple of terms before I should have done. I had been kept home to make up the required number for the threshing gang—it was not always easy to find nine helpers—and because of the priority given to food production, farmers were allowed to keep their sons from school for urgent work. My job at that particular threshing was to cut the bonds; and I remember to this day overhearing one of my father's regular workers say: 'He's still at school and his hands are soft.' I felt ashamed.

On my return to school, after an absence of a week or ten days, my form master said that as I was in the habit of stopping at home to work, he did not see much point in my remaining at school. His opinion coincided with mine; in fact I had formed it some time before, and I dare say he had, too, for I had been fairly successful at resisting all attempts to educate me. I had no reason to suppose my success had gone unnoticed.

And so it was that I left school, and started to harden my hands.

# Making merrie in a wolfish month

## January 1983

The middle of winter seems an odd time to start the year; Michaelmas or Lady Day would be far more appropriate. In fact Lady Day, March 25, was the official beginning of the English year from the twelfth to the eighteenth century, and it was not until England adopted the Gregorian Calendar in 1752 that we returned to the ancient new year of January 1.

The Anglo-Saxons called the month Wulfmonath, it being the time when wolves starved and descended upon their villages. The wolf was not exterminated in Britain until the mid-eighteenth century, but I dare say that if there were still a few here today there would be a movement to save them; for all I know, there may even be one to bring them back. In other ways, January can still be a 'wolfish' month, with the eerie nocturnal mating cries of the fox a reminder of the way our ancestors must have felt as they huddled in their villages and heard the howling of prowling wolves.

When January was introduced into the Roman calendar it was named after Janus, the god of doors and the god with two faces. Like Janus, we too look two ways at this time of year. Our glance back to the old is one of some affection and regret, no matter what kind of year it has been; after all, we lived with it for twelve months and had begun to get used to it, while the new one is unknown and unpredictable, an upstart pushing aside an old friend.

In Gloucestershire on Twelfth Night, thirteen fires used to be lit in honour of Christ and his disciples; the fire named after Judas was stamped out immediately while the others were allowed to burn out.

In orchards there was wassailing to encourage good apple crops, with cider poured round the roots of a favoured tree and the song:

> Here's to thee, old apple tree,
> Whence thou mayest bud, and whence thou mayest blow,
> Whence thou mayest bear apples now!
> Hats full! Caps full!
> Three score bushels full!

Traditionally, the coldest weather of the year is supposed to fall on the day of St Hilary, January 13, but the coldest recorded day was January 20, 1838, when the temperature fell to minus 14°F. Last year our coldest day was January 14, with the temperature down to minus 8°F. That was the day I walked by the Severn and watched the ice floes moving slowly downstream. The river was even greyer than the sky, and sinister too, and as it moved it hissed like some huge angry serpent.

For days we had lived in a hushed world of white, the fields covered with eighteen inches of snow, great canopies of it overlapping the roof tops; swirls and drifts of snow, icicles hanging from eaves, frozen ponds, a leaden sky laden with more snow. No vehicles, but plenty of people on foot, children with toboggans, and cattle with icicles hanging from their coats which tinkled as they moved. Stark black trees, frozen sunlight and above all that, a breathtakingly beautiful but grim and foreboding-laden sky, a scene such as Breughel might have painted.

We were all bound by snow, if not exactly snowbound, thrown back on our own resources, beset with difficulties; even walking was difficult enough. We could almost have been in medieval England, and yet in spite—or because—of all the difficulties, the inconveniences, the coldness almost beyond the point of endurance, everybody seemed to have more time; time to talk, to smile, to be more cheerful and friendly.

Perhaps Merrie England was more merry than we often imagine.

# Mangolds on my mind ...

The first regular piece I wrote for this magazine was about mangolds, and I have made other references to them since. An old lady in the village who reads these columns and has known me since childhood was reported as saying: 'I believe he intends to grow mangolds. Oh dear, all that mud.'

I had intended to grow mangolds for some years, though I confess the thought of all that mud had deterred me, too. But they stayed on my mind, a token of a return to a former husbandry, and I was haunted by memories of the sound and delight of cows eating them. Yes, I *would* grow mangolds again.

A neighbour cautioned me: 'A lot of work, and in a wet autumn all that mud; it isn't as if you've a horse and cart.'

But I would not heed caution. We prepared the ground, sprayed it against weeds and asked a contractor to come and sow the seed with a precision drill to avoid the laborious task of hand-hoeing. Traditionally, mangold seeds were planted in clusters, which used to mean a lot of singling, but now rubbed or separated seed is obtainable.

I bought a variety called Wintergold, a happy choice of name, and worried a little when the merchant told me that a good deal of mangold seed had been sold that season. Had I dallied too long? Could I be in danger of being fashionable and—perish the thought—trendy?

That sceptical neighbour happened to be passing just as the contractor was drawing into the field, and gave us a most peculiar grin; amused, cynical, envious, mocking or scornful, I wasn't sure.

The burly, bearded driver drilled kale first, a few rows of swedes and then the mangolds, all on a day that reminded me of Edward Thomas:

> It was a perfect day
> For sowing; just
> As sweet and dry was the ground
> As tobacco dust.

The month of May, you will recall, was exceptionally dry. A squall or two in early June started the kale and swedes, but not the mangolds: when they did appear they were very sparse, not nearly so vigorous as the weeds. I thought of ploughing the whole lot in. Instead, I took a hoe ...

There was a time when I wondered if I was wasting my time, but day by day more seedlings appeared. There were doubles, as the bearded man had warned me there would be with rubbed seed—get pelleted seed next year, he had advised—but as the plants were so weak I left them as they were.

Nicholas and Rupert spent a few hours hoeing beside me—Simon being otherwise engaged—but most of the time I was alone in the field. The longer I hoed, the more I enjoyed it. With the quietness and the solitude, my mind became carefree, and I belonged to a world of bliss, just hoeing with bent back. Even the ensuing tiredness and aching back were a kind of benison.

This, I thought, was a man's natural order and pace; one could meditate or let the mind go blank—but you have to be middle-aged to think like that, and perhaps a bit of a dreamer, too.

By October the mangolds had become a most respectable size and, being late starters, were still growing. That month was wet, very wet, and I began to have doubts about hauling them off the field. All that mud.

I saw my neighbour, and he spoke dolefully about the weather and the mud and what a bad time it would be for anyone with mangolds to harvest.

November came in fine, the mud began to dry, and we started pulling. The old rhythm returned; grasp the leaves in one hand, slash through them with a knife, and the root tumbles off on to the heap. I did not lack volunteers, young and eager helpers: Rupert, Simon, a friend of each and, one afternoon, Rupert's wife as well. Rather surprisingly, they enjoyed the work as much as I did.

Next we loaded the mangolds on tractor trailers and hauled them to the clamp, all the while free of 'all that mud.' We reckoned we had forty tons of mangolds to the acre; my neighbour cast his eye over a load of them and declared them 'whoppers;' and the clamp was well covered with straw well before the first frost.

Now it is January, the mangolds have matured, and the best part of the story is yet to come. For as I told you before, through those long mangold-less years, I was haunted by memories of the sound and delight of cows eating them ...

# Bread

In the depths of one of the first of this winter's series of strikes, I thought how sad it was to see the queues outside the independent bakers' shops. Sad because these private bakeries, like other small tradesmen, are so few; so many of them have been forced out of business, yet if we want service and quality, it is the small businesses that provide them. Sad, too, because so many housewives have lost, or never acquired, the art of baking bread, having believed the nonsense that it is difficult.

But there is so much nonsense talked about bread. Nonsense to call that mass produced stuff bread. Nonsense to eat that insipid white pulp with most of the goodness removed from it. The fashion started from snobbery; once, only the rich ate white bread, so, when they could, the poor began to do the same, and scorned the dark wholemeal loaf. It has now turned full circle; I suppose it is 'posher' to eat wholemeal bread today. But 'posher' or not, it is more sensible to eat the real stuff, the stuff of life.

The greatest nonsense is that English wheat will not make good bread. My wife has baked our bread for years, and then, a few years ago, we became tired of paying four times as much for a hundred-weight of flour as we received for a hundredweight of our wheat. So we bought a hand mill, and started grinding our own wheat. It is, it must be admitted, a slow process, and jolly hard work. When it is my turn to grind the flour, I comfort myself with the thought: 'By the sweat of thy face shalt thou eat bread.'

At first, the result of baking bread from our own wheat was a soggy disaster. Then my wife reduced the amount of water by almost half;

the outcome was a delicious, nutty-flavoured bread, real bread. I began to see why the myth that English wheat would not make good bread was created. It is not bad business if you can sell the public water—or is it? I still fail to see how the myth was established. Surely the English had made bread from English wheat for centuries?

Of course, there were powerful reasons for creating the myth. The Industrial Revolution had meant that we had hardware to sell to a world eager to buy it. But we needed paying for the hardware, so the buyers sold us cheap food, impoverishing themselves, their land, and English farming. And many of our experts are still living in those days, it seems, their thinking based on the panacea of more and more production of hardware or the like—and on cheap food.

Cheap imported food destroyed our corn mills, the focal point of country life, and English farming was allowed to dwindle. The U boats of the First World War restored English farming, but memories were short, and it was allowed to deteriorate until the next war. Afterwards, a system of subsidies was introduced, because food had to be cheap. Then, by a combination of many factors, men were lured or driven from the land, often in the name of efficiency.

It does seem odd to me that although food is man's prime necessity, it is deemed needful to have fewer and fewer workers engaged in the one line of production that is not a false panacea. And when men are driven from the land to join the dole queues in the town, madness seems the only word to describe it. All of which returns me to my original subject, queues.

# A landscape of leafless gloom

## January 1984

Unless he has some definite objective a farmer does not go walking; he will walk to inspect fields, crops or livestock, but will never just walk.

It must have been the exceptional autumnal air and scents, the turning leaves and the sunlight that tempted my wife and I to take to strolling on Sunday mornings a few weeks ago. That, and because it was so dry underfoot. How rare it is for the land to be as dry as that in November, and how rare, indeed, to have such a fine, glorious autumn generally. I cannot remember another so splendid, but memories are as short with the good weather as they are long with the bad—the hard times we cursed while we were experiencing them, but boast about in after years. How we English grumble about our weather, but how we boast about its inclemency.

Above all else it was the sun that persuaded us to go walking. 'There's nothing like the sun as the year dies,' said Edward Thomas. 'There's nothing like the sun until we are dead.'

Along the road by the tall hedge with bright hips and spindle berries, festooned with clematis—old man's beard—now turning grey; beyond the alderlined stream, willow trees like silver billows merging with the sky.

Such an abundance of grass, and so late in the year; cattle still outside and leaving hardly a hoofmark, and autumn-sown barley and wheat growing as if spring has already arrived. A blackbird sings in a hedge, a green woodpecker, resenting our presence, flees indignantly away.

Into the wood, along a path strewn with large-leaved, earth-hugging brambles, and on either side oak and ash. The brown oak leaves still cling firmly, but the ash lie flat and green on the ground. Leaves tumble gently from chestnut and lime, birch and hazel; and crisp and curled, brown, bronze, tawny, gold or lemon, they rustle against our feet.

Under the freshly fallen leaves is a soft carpet made up of their predecessors of former years, while from above comes the benison of sunlight. Slowly we make our way on this morning made for taking life slowly.

Then the scene and the mood change; the happy but haphazard hardwoods are replaced by serried ranks of conifers; gaiety gives way to gloom, and variety to uniformity. There is no galaxy of colour beneath our feet, no leaves drift through the sunlight; our light and airy English woodland has become a dark, forbidding Teutonic forest.

During the week I had been working near another woodland. Around its perimeter there were oak, ash, alder, holly and hazel still, but beyond stood the massed army of conifers. This, too, until quite recently, had been an ancient woodland. Ecologically, these dense stands of conifer are an offence against the landscape. Whatever arguments are invented, it still remains 'a sin against the light' to introduce these ranks of softwoods in a landscape where hardwoods are indigenous. In the Forest of Dean they stand behind hardwood screens like alien armies hiding.

According to a census published by the Forestry Commission there has been a considerable loss of oak woodland, as much as 150,000 acres, during the past thirty years. The 40 per cent increase in woodland has been almost entirely due to the planting of conifers, although the commission claims that oak is still the most dominant species in England. It is certainly difficult to accept its finding that there has not been a decline in the total area of broadleaved woodlands, but an 11 per cent increase—especially if you have looked beyond the screens and seen the softwoods where hardwoods once stood. Equally surprising is the claim that, in spite of the loss of the elms through disease and of hardwoods through the removal of hedges, there has been a 15 per cent increase in non-woodland trees since 1951.

This census took three years to complete, and is the most detailed one for almost forty years. Possibly only the Forestry Commission has the resources to conduct such a survey, so it is difficult to challenge its accuracy. Its findings must strain the resources of even the most credulous. The rest of us have only the evidence of our own eyes, and it tells us that the Forestry Commission, which now controls 9 per cent of all land in the United Kingdom, is without doubt altering the character of the English countryside.

# On foggy days you can see forever ...

January 1985

'Be Suer of hay, till th' end of May,' advised old Tusser, and it is still sound advice, even after four hundred years. All spring and summer we were growing and gathering crops against the siege of winter. By September the barns were full of hay, corn and straw—but a full barn will wax hollow, and by November the signs were there. Now, at the start of January—and with another month before we reach the middle of the winter feeding period—those stacks have dwindled alarmingly.

Except the mangolds; that store is as yet untouched, mangolds not being suitable for cattle until Christmas is past. It will stay unopened and untouched until tomorrow or the day after, this store of winter wealth.

The mangolds were the last crop to be harvested. It was on a day in mid-November when we pulled the last of them. We were working close to the stream lined with alders, which we call arle. According to Virgil, the world's first boats were made of alder wood. More recently the wood was used for clog-making and turnery work—a wood which is white when freshly cut, but which soon turns an attractive reddish colour.

Some of the dark green leaves of alder had already fallen as we twisted the fleshy leaves off the mangolds. It was foggy on that mid-November day. We could see no farther than twenty yards, if that. We worked in a little enclosed world, unseen and unheard, far from the noise of men and machines.

On the bank of the stream stood a solitary plant of honesty (*Lunaria*), its white stems and silver moons bright against the fog.

In this hushed world even the stream was silent, as quiet as a nun. But this stream, so gentle today, can on occasion become a raging demon, running amuck over the ground on which we work. These floods often bring plants which establish themselves on the bank; the honesty was one example, and balsam is another.

The fog and our restricted vision, the quietness and the peace of our world and of our work gave us a sense of timelessness; if time had not ceased to exist, it had at least eased its headlong rush. Our work, the quiet, slow manual labour, was timeless, too. We could have been medieval peasants, although English ones would not have pulled mangolds, since they were not introduced into this country until the last century.

We were working as generations before us had worked, and as generations to come may yet work. This craze for bigger and more powerful and more expensive machines, more noise and more speed, may not last for ever. Already some of the ways and wisdom of our forefathers, so recently derided, are being rediscovered.

My left hand grasped the leaves. My right hand, which held the mangold, was covered in mud. Simon and Rupert wore gloves, of which I secretly disapproved. I liked the feel of the mud, the earth which had grown these mangolds and sustained generations before us, just as those gracious alders, with half their roots in water, had sustained the banks of the stream against the ravages of flood.

The past seemed so close to the present. The fog seemed not oppressive but protective, upholding that illusion of time standing still, the past and the present merged into one.

It was a dull day, to be sure. Our work was slow and laborious, and some may deem it drudgery. Yet to us it wasn't like that at all; there is something peculiarly satisfying about pulling mangolds, as there is with digging, hoeing, and so much else when you have direct contact with the earth; the earth to which all of us belong, whether we realise it or not.

Perhaps one needs to be a poet, a romantic, a peasant or just simple-minded to find pleasure and contentment in such a way on such a day. All I know is that I harvested more than mangolds that mid-November afternoon.

# Of flagstones and flagons ...

The carpet on the floor of the country pub is a clue to the change in rural areas, but there are still a few houses that welcome the dwindling number of men with mud on their boots. The Red Hart at Blaisdon has kept its bare-flagged floor, countrymen and country atmosphere, and its name, too, makes it distinctive; there are plenty of Red Lions and White Harts but a Red Hart is a rarity.

The name seems deeply linked with the Forest, for there is another Red Hart at Awre. I have certainly never heard of any others in England, although I believe there is one in Wales.

The Red Hart used to be the only place in Blaisdon which did not belong to the estate. One squire tried to buy it, and when he was unsuccessful he set out to get it closed. It survived, as it still does, surprisingly, perhaps, since so many country pubs have closed, and Blaisdon is such a small village, too. The Hart has always attracted countrymen from a wide area, robustious men who gave character to the place.

There was Harold H., who took his horse into the bar, flung a fox in there, threw a bottle of whisky in the fire, and poured a pint of beer over anyone who refused a drink. The recently-deceased and much-lamented Albert metaphorically lived there for years, and literally lived there for the last two years of his life. Albert knew as well as Dr Johnson that nothing contrived by man produces so much happiness as a good tavern or inn.

For almost thirty years, until his death in the early 1950s, Gideon Price was the landlord; the pub is still in the same family. In the old days there were only two rooms, the bar and another which was rarely used. The beer was drawn straight from wooden barrels, and cribbage, quoits and shove-halfpenny were the games; there were no darts flying about.

Jars of Bluett's animal ointment stood on the shelves, a copy of Old Moore's Almanack hung on the wall, and over the fireplace was a picture of the Mitcheldean Brewery, which once owned the pub. Behind this was the butcher's saucepan, into which he poured his beer to be warmed on the fire on winter nights.

Gid, with his beard, cap, cardigan, and breeches undone below the knee, would sit by the fire and discourse—about pigs, plums, the weather, cabbages and many things. None but Harold H. dared contradict him.

As I write I realise, with some surprise, that I have been going to the Hart for almost forty years. There was a time when, with the naivety of youth, I thought those characters would last for ever, such was the enduring solidarity they had about them; that the beer would ever be drawn from the wood; that Old Moore would always be consulted and Bluett's ointment in demand; and that I would always be young.

Those robustious characters have gone and so has my youth, for now, like they once did, I start so many sentences with 'I remember.' Old men forget, but the older they get the more they remember, and memories get bigger and better with age.

Yes, I remember the tales and the snows and the men of yesteryear, strong Cheltenham beer and those flagons of stunum which made a man lose his legs and discretion.

There have been chances at the Hart. The bar room has been enlarged, and now there are two others, one a lounge and the other a pool room. The picture of Mitcheldean Brewery has gone, as have Old Moore and the Bluett's ointment. Beer is no longer drawn from the wood, but through an array of gadgets on the counter, top pressure and keg and lager. And though there is still Guinness in the bottle, thankfully, gone are the flagons of stunum.

Darts are more popular than cribbage, quoits or shove halfpenny, though Gid always said the ceiling was too low for darts. In fact I suspect he forbade them for much the same reason that Henry VIII forbade shove halfpenny; others played better.

Yes, the Red Hart has changed—even the lavatories have improved—but in essence it stays the same; that bare flagged floor is more than a symbol or a talisman, for as in all good pubs, it is a sign that democracy reigns supreme, locals mix with newcomers, and artificial barriers are broken by talk, laughter and song.

Hubert will sing the story of Colin and his cow, lead the company in *A Farmer's Boy*, and sing racy songs when he is mellow. Sometimes the ghosts seem to outnumber the living, but I fancy they are talking and laughing and singing, too. If we lose inns like the Red Hart we might as well join them in their other abode, wherever that may be. As Belloc said: 'When you have lost your inns, drown your empty selves, for you will have lost the last of England.'

# Going, going, gone!

## February 1978

Many farmers will have affectionate memories of the old cattle market beneath the trees at Gloucester. I call it the old market but, of course, it was once the new one. Until a market place was established in 1821, livestock sales were conducted in the streets; Barton Fair, the annual sheep and horse sale, is a reminder of these times. In 1855 a new and larger market was built, and this was further enlarged, in 1874, for the sale of poultry, fruit and vegetables.

The market was on the site now occupied by Grosvenor House and the bus station, right in the town and convenient for the GWR and LMS railway stations. The railway played an important part in the life of the market right up to the latter's closure in 1960. And so did the several public houses around the market; The Lamb, The Crown, The Prince Albert, The Wellington, The Gloucester, The Spread Eagle, and others. Most of them, like the railway stations and the market itself, are gone. Going, going, gone, as the auctioneers cried so many times.

And gone with them is an era, an atmosphere, an indefinable something; farmers over fifty will know what I mean. On market days the country invaded and pervaded the town. Farmers, cattle lorries and animals filled the market, and the streets around. Farmers' wives came for their shopping. The old market had a warmth, sadly lacking in the new one.

Gloucester was a market town, then, and had vigour and character. But of course, it couldn't continue. The market was lively, colourful, noisy, and a great nuisance. Another source of annoyance to the authorities was the exemption of market tolls granted to all farms of the Duchy of Lancaster; there were several in the parishes of Minster-worth and Westbury-on-Severn, for instance.

So a big new market was constructed just outside the town, farms were deprived of their exemption, Gloucester was no longer cluttered up with the country, and really ceased to be a market town. Of course, the old market was far too small. It couldn't cope with present day trade, the smaller markets, for one reason or another, having been slowly killed.

The new market is so much more convenient, but it's so much duller, colder and more impersonal. So big that you can wander about there all day and not find whoever you're seeking. As is the way with progress, when you stop to think, the losses so often outnumber the gains.

# Old days of hay

My companion pointed to the straight quickthorn hedge and said: 'I helped plant that when I was a boy. It's a mile and a quarter long.' We were at Corn Ham, on that level tract of land at Minsterworth, extending to almost a thousand acres and reaching to the bank of the Severn.

'That's good old ground, that'll grow whate an' byuns,' an old Minsterworth man once told me, but my companion this afternoon was more interested in the hay that had been produced there.

'Aye, and think, it all had to be cut with scythes, once,' he remarked. The thought was daunting. The task seemed insurmountable, but the answer is given in those old photographs of ranks of men armed with scythes. Even so, it must have been hard work; muscle, skill and sweat. Only, forty years ago haymaking was still dependent on the muscle of man and horse.

We were only a few score yards from the river when he said: 'That's where the ricks were built.' I saw the staddles of earth and stone, about sixty by forty feet and almost four feet high. 'To kip 'em out of the flood,' he explained. 'The hay was cut and trussed in presses and loaded on the longboats.'

He climbed over the sea-wall and made for the river's edge with a speed surprising for his age. 'See here,' he said as we stood on the river bank. 'There's the remains of a wharf. Then the hay would go off to the docks at Gloucester.'

'And where would it go from there?' I asked.

'Oh, I'd aim it 'ould go off to London or somewhere.'

Having seen the hay to Gloucester Docks, so to speak, he seemed to lose interest in it.

Those were the days when the whole country moved on hay, the 'mother of the horse.' At one time there were fears that there wouldn't be enough—a fuel crisis, we'd call it today. The towns used vast quantities. Everything depended on hay, not just agriculture, but transport, commerce, industry. Even that alternative source of power, coal, was dependent on hay for the pit ponies.

Earlier the canals and the railways too had needed horses, and consequently hay, to set them on their way. The First World War was fought on hay; the Government commandeered stacks of it and shipped it off to France. Hay is romance and history, and even in present day farming with silage, kale and compound feeding stuffs, 'good hay', as always, 'hath no fellow.'

# Deliberations on a lengthening day

## February 1980

Lengthening days and the appearance of snowdrops can give the illusion that spring is just around the corner. But remember the old saying: *As the day lengthens, the cold strengthens.* These are the hungry months. *From Christmas to May weak cattle decay;* and *A late spring never deceives.*

There is, to use a phrase of Dr Johnson's, a bottom of good sense in many of the old country sayings, and they often illustrate how little the fundamentals of farming have changed.

*Take half your straw and half your hay on Candlemas Day.* The bright young men from the Ministry, the fertiliser manufacturers and other 'experts' would have us believe that cattle can be grazing an abundance of grass in March. So they might—about one year in ten, if the land is dry enough to prevent their having *Five mouths instead of one.*

Candlemas Day, February 2, is still only half way through the period of winter-feeding, and the prudent farmer will indeed have half his stock of winter fodder still. But how many farmers *are* still prudent? Too many have been seduced by the blandishments and the glib sales-talk of the 'advisers,' and their farms are overstocked. My vet is of the opinion that many farmers could increase their profits, and reduce their problems and their overdrafts, too, simply by reducing their stocking rate.

If I send a barren cow to market, I may be given a good price for her—and an uneasy conscience, because she could well be exported live for slaughter. The export of live animals, with its attendant cruelties, is a scandal, a blight on agriculture and the nation; yet it still continues.

Many vets and farmers deplore the trade, and it is morally and economically wrong; but the National Farmers' Union supports it.

Plymouth Corporation is attempting to have the trade banned from its docks—and the NFU intends to fight the move. Once again it will put all farmers in a bad light; its past record shows that it is always quick to defend intensive factory farming, but not too keen to uphold worthier causes.

The RSPCA is campaigning to prevent the unnecessary ands repulsive cruelty to animals caused by intensive methods, degrading to both humans and animals; I dare say the NFU will do its utmost to defend them—and so, lacking any coherent long-term policy for agriculture, the union staggers from folly to folly and agriculture stumbles from crisis to crisis, falling further into disgrace.

Devonshire, Gloucestershire, Herefordshire and Somerset are, or rather were, the cider counties. William Marshall, in his *Rural Economy of Gloucestershire* of 1789, said that farm wages were very low in money but shamefully exorbitant in drink; six quarts of cider was the daily allowance, frequently two gallons, sometimes unlimited. Drinking a gallon bottle at a single draught was no uncommon feat, a mere boyish trick. One man of the Vale, to be even with his master who had paid him short in money, downed a two-gallon bottle without taking it from his lips. Marshall saw this pattern of payment as a fault and a crime.

Cider continued to play a large part in Gloucestershire agriculture well into this century. Until forty or fifty years ago most farms in the Vale made their own, and even twenty-five years ago there would be little extra help forthcoming at hay and harvest times unless the farmer had a stock of drink.

Now very few farms produce the genuine cider any more than real farm cheese, butter or bacon, for there has been a general decline in self-sufficiency. Before we grow too nostalgic about the good old rough, potent, thirst-slaking farm cider, perhaps we should remember Marshall's account; still it would be nice to have a tot or two—and maybe a few more.

A Museum of Cider has been established at Hereford; and recently it has published a booklet, *A Drink For Its Time*; an account of cider making in the Western Counties by Michael B. Quinion. Well illustrated with many splendid old photographs, it is a bargain at 60p.

Changing times and better wages put an end to farm cider; changing farmers, too. My grandfather went on making cider until he died at the age of 92; and would still be making it today if he were alive. There is another reason, as a farmer told the booklet's author: 'Well, I paid 'em to make it, and I paid 'em to drink it, and still the buggers weren't satisfied. So I stopped making it.'

I still know a few places where they make the real stuff, nothing like the drink sold over the counter. And you can still buy some tidy pints in Gloucester on market days.

Marshall's *Rural Economy*, incidentally, was printed by R. Raikes of Gloucester. Nine years earlier Raikes had started his Sunday schools, of which this, of course, is the bicentenary year; Catherine Boevey of Flaxley Abbey had been doing much same thing a century earlier, and hers is a fascinating story, too. No archetypal schoolmarm, she was married at fifteen, widowed seven years later and famed throughout England for her beauty, talent and charity. A friend of Pope, Steele and Addison, she was immortalised in *Spectator* essays as the 'Perverse Widow' hopelessly pursued by Sir Roger de Coverley.

# When men could keep
# a firm grip on life

February 1985

A crack appeared in the handle of my spade last spring, but bound with tape, it served, and I dug post-holes and the garden again in the autumn. Then, one day this winter, I took the spade to dig some trenches on the headland of a corn field where the water was hanging. Winter corn will tolerate wind and frost, and snow will protect it like a blanket; but it does not like stagnant water.

It was a pleasant, rewarding task, digging those narrow trenches to release the water, and seeing it gushing into the ditch. I was absorbed in my work—work ceases to be labour when it absorbs you—but then I heard an ominous sound, and found that the handle of the spade had parted from its blade.

It had taken me some time to become accustomed to that spade and its handle. At first the handle seemed too short; Gloucestershire spade handles used to be longer than most, but now they are no more, for spade handles, like so much else, have become standardised.

Fortunately I knew a man whose hobby is fitting new handles to tools, and soon my spade was mended, returned with handle and blade carefully oiled. I mentioned the Gloucestershire handle; yes, I was told, they were longer, and with the real old Gloucester the handle used to fit right down into the top of the blade, too.

How important handles once were, and how much of a countryman's life was attached to them. Handles of tools, plough and horsehoe handles, handles of root-pulpers, cake-crushers, chaff-cutters, handles worn thin and shining with constant use. Then came another handle, a metal one, the forerunner of what was to put so many of those wooden handles out of use—the tractor-starting handle, which has itself grown obsolete.

Those hand tools with their wooden handles were cherished and guarded by their users.

The hook, the hoe, the scythe—especially the scythe—the bill-hook, hedge-bill, axe, spade, even the fork. In the hands of a skilled man they became animate, an extension of his limbs. 'These men trust to their hands; and every one is wise in his work;' for two thousand years or more those words in the Bible could stand as they were written. Now the past tense must be used—and two thousand years of experience have been exchanged for what?

The gain, if such it be, is speed. Unless we do something quickly we feel we have lost something. That something is time. Time is a most precious thing, but when we have gained it we are then at pains to kill it.

George Bourne (in *Lucy Bettesworth*, 1913) recorded two beliefs about scythes, the first being that all scythe blades were stamped with the day of the week upon which they were made. Those made on Wednesday, Thursday and Friday were considered superior because on those days the heat of the cutler's fire was at its best. The Thursday scythe was the most prized—and was this, I wonder, because Thursday was Thor's day? Thor, the god of peasants, the son of Woden, the god of wisdom, poetry, war and agriculture ...

I personally doubt that scythes were ever stamped in the way Bourne suggests, but myths can be more powerful than facts, and I dare say a man who thought he had a Thursday's scythe did find it a better tool.

The other belief was that if a scythe was left hanging in a tree, a better edge could be obtained. 'It rusts the iron out of them,' Bourne was told. That was in Surrey, but later, in Essex, Henry Warren heard: 'That rusts 'em, and then you can get a better edge.'

Tolstoy used to use a scythe, mowing with Russian peasants. In his *Anna Karenina* his character Levin does the same, and at the end of the day reflects: 'What a remedy it is for every kind of folly.' Levin's remark could also be applied to the use of other tools; and so it is that Tolstoy, not noted for his brevity, sums up in a few words what I have tried to say in several.

# Curious cows, docile dogs

February 1983

Anyone who has ever sat in a field where there has been a herd of cattle will have discovered how inquisitive the beasts are. For some reason there is nothing that arouses the curiosity of cows more than a human sitting in a field.

I know a farmer who inspects his cattle during the summer by calling them, sitting down and waiting for them to come to him. His fields are large and as he says, it saves him a lot of walking. Of course some cattle are more inquisitive than others, just as some appear more intelligent than others. But I wonder whether humans can really judge the intelligence of animals—and by the same token, are we right in attributing certain characteristics to them, or in applying those supposed characteristics to our own species?

Sheep are not credited with much intelligence or sense—not that intelligence and sense are synonymous. Yet sheep can usually find a means of escape when they wish to move to fresh pastures, and compared with humans they are sensible and rational beings. Sheep are never 'sheepish', meaning shy or abashed in the way that a person can be.

On the other hand, bees are regarded as most intelligent creatures—but move their hives a few yards, and they are lost. They are certainly industrious, making countless heavily-burdened journeys all summer long to make honey they will never taste; in comparison wasps are indolent, feckless fellows, but at least they know how to enjoy their brief span of life.

We tend to measure the intelligence of animals by the use we can make of them, by their likeness to ourselves, by their ability to learn the tricks we teach them, or by the standards we use to gauge human intelligence. Surely, a really intelligent cow is intelligent in a bovine way—a way that could be worlds apart from the intelligence and ways of a human.

A dog, so long domesticated and used as a companion of man, has learned man's ways and tricks, does his bidding, and becomes dependent on him. More, it has become subservient and fawns on man—which means that man, arrogant and susceptible to flattery, rates dog very highly indeed. Is it at the loss of dog's natural intelligence, though? It is a pity that we cannot ask the opinion of the fox.

The cat, on the other hand, has been domesticated for thousands of years, but it has not lost its native intelligence—or its pride. You may have a cat about the place, you may even call it your cat, but it remains its own master. You cannot own a cat, although it may own you, consenting to live with you on its own terms. It may play with you, if it amuses it to do so, but it will not do foolish human tricks to amuse you, and if somebody writes to say that they know a cat that will, I will reply that it is just a silly cat, the exception that proves the rule.

All this talk of intelligence in animals is just poppycock; we cannot prove or judge another species' intelligence. I am not even sure that animals need it, anyway. Man has intelligence, animals have sense; and while intelligence may destroy the world, only sense can save it.

# Wonder of whitethorn

February 1984

There are hedges and hedges—and hedges that are no more. Whitethorn, also called hawthorn or quick, makes the best hedge, and well-maintained, it is stockproof and needs no assistance of barbed wire. Most all-whitethorn hedges are straight; they have usually been planted, and generally they date from the time of the enclosures.

Winding hedges nearly always contain several species: blackthorn, dogrose, elder, hazel, holly, maple—they vary with the soil and the district. Ash and elm are also common, but both are poor hedging material, and bramble and clematis both tend to smother a hedge.

Hedges composed of many species are probably very old, dating from before the times of the enclosures. It is a fallacy to suppose that there were no, or very few, hedges before the enclosures, but a useful fallacy for those who seek to justify the grubbing of hedges.

It has been suggested and is now widely accepted that the age of a hedge can be estimated by the number of different species in a certain length—thirty yards, I believe; each species, it is said, counts for a hundred years, so a hedge with eight different species within the specified length is eight hundred years old.

This system would prove the widespread existence of hedges before the times of large-scale enclosure, if it were accurate, but it also presupposes that even these hedges were deliberately planted, and that once every hundred years another species crept in. I find this method too pat, too simple, the work of the academic rather than the practical man.

Much more likely, our far-off ancestors made use of the species already growing on the sites where they decided to make their hedges.

That would be the way of practical countrymen, to make the best of that which was already there.

Just as there are hedges and hedges, so there are, or were, hedge-layers and hedge-layers. Good hedge-laying is an art, and the number of artists was always, I suspect, comparatively small; today such craftsmen are very rare indeed.

The style and method of laying varied from district to district. In the Berkeley Vale stakes were not used, but in most districts they were an integral feature. The hedge-layer's equipment consisted of an axe, billhook, hedgebill, stake beetle, fork, and a pair of horsehide gloves or mittens, each with two divisions only, one for the fingers and one for the thumb. An iron bar was also used sometimes to make holes for the stakes.

The stake beetle was a home-made wooden club, with a hollow to take the chamfered stake to prevent splitting, or leaving the stake looking like a shaving-brush. Only the bad workman used an axe or mallet to drive stakes, and the good workman scorned the use of a saw.

The hedging bush was pleached or plaited low, and the projecting snags were cut away. The pleaches were never bent or woven round the stakes, and they were laid so that all their tops were on the same side of the hedge. The stakes were driven in at a slight angle, so that their tops inclined towards the part of the hedge yet to be laid. Stakes in my district were of willow, for a long-lasting one was neither needed nor desirable. Live stakes—material already growing in the hedge—were anathema to the good hedge-layer.

If the hedge was to be ethered, or bound along the top, the stakes were vertical, and long rods, usually of hazel, were woven along the top of them, a rod to every stake. The finished hedge was not tightly-packed; it used to be said that a blackbird should be able to fly through one.

Such hedge-laying was an art, beauty and utility combined. Today few hedges are laid, and still fewer laid to perfection; indeed many of today's best efforts would have brought snorts of sarcastic scorn from the hedge-layer with whom I worked. My job was to rid the hedge for him, cutting out all unwanted growth; to cut, sharpen, chamfer and carry his stakes; and to be his general factotum. Myself, I never properly mastered the art of good hedge-laying; but at least I learned how it should be done.

# Room for improvement

## February 1979

Something must be done. My wife had been saying it for some time, but now she said it in a tone which meant that something *would* be done—and done immediately. She was referring to the state of my room, where I do my writing and accounts, and where I swear about forms and VAT.

My room faces east; the morning sun comes slanting in, over the orchard, across the garden, through the window and on to my desk. It would be pleasant sitting at my desk in the sunlight, were it not for the guilty feeling I always have; I know very well that fields are the proper place for a farmer when the sun is shining.

For weeks my room had been in a muddle. The desk littered with papers, accounts, unanswered letters—even unopened ones; and somewhere among the mess lurked some cheques I hadn't paid into the bank. There were catalogues, pens, empty tobacco tins, books, and a lot more besides. Newspapers, magazines, more books and catalogues and papers in the chairs and on the floor, and lots of cardboard boxes holding more papers, VAT forms and invoices.

'Look at it,' said my wife. 'It's a disgrace. I can't get in to clean this room. It must be tidied up this morning.'

'Not this morning,' I said. 'Not while the sun is shining—a wet morning perhaps;' anything to delay a task I hate.

It's the same outside. There are so many jobs we put off for a rainy day, and the trouble is there are not enough rainy days, or if there are then we discover that many of those rainy day jobs require fine weather, after all.

Of course the real trouble is that I'm a hoarder. Take those tobacco tins, so nice and so useless at the moment, but one never knows, they may come in handy one day, to quote those fatal words. Apart from which, my method of tidying up—and, I suspect, a great many other people's—is simply a case of moving a muddle from one corner to another. And of course there's always the danger of destroying something inadvertently.

On the farm there are old bits of iron, short lengths of wood, all manner of things which may be very useful one day—so they're put aside, and another muddle is in the making.

Of course my wife is right, one must be ruthless; rubbish must be thrown away. But what is rubbish? I've lived long enough to see a lot of rubbish thrown away, and to discover it wasn't rubbish after all. Old farm wagons and carts, harness, tools, implements, winnowing machines, butter churns, cider presses, all no longer in use, just cluttering up the place, destroyed, sold or thrown away. And look now, collector's items, much sought and prized—and worth money, too.

It makes me wonder if what we're throwing away today will be cherished in the future, too: old oil drums, rubber tyres, plastic this and plastic that. No, never, never—and yet it's a condemnation of our times, isn't it? You can't be sure; the evidence is with us hoarders: 'How much am I bid for this plastic fertilizer bag, circa 1979?'

My wife didn't even listen. 'Those farm magazines, they're last year's, you don't need those.' And yet only a few weeks ago I paid fifty pence per copy for some 1920s *Farmer and Stockbreeder*, cover price two old pennies.

'What's in those boxes?' she demanded. I wasn't quite sure, but I thought it had better be kept. The tobacco tins and those large used envelopes disappeared when I wasn't looking.

At last order was restored to my room; books replaced on shelves, pens and paper to their proper places, invoices and other papers filed away, stamps and cheques found. Really, the room looked fine, and a man can settle down to write on a clear desk; I always say that there's nothing like a good tidy-up to make everything look better and brighter.

Now I could write this column in comfort and with a clear conscience, which I couldn't before. But first I hung the large framed photograph of my grandfather on the wall. It was taken about forty years ago, when he was in his eighty-fifth year. For another half-dozen years after that he was still active, spent almost every daylight hour working on his farm and climbed the steep hills every day to see his beloved sheep.

On the farm he always wore a thick white smock; a bowler hat green with age in winter, a straw hat during the summer. But in this photograph he's hatless, his thick white hair carefully brushed and wearing his best clothes, his swallow-tailed jacket unbuttoned to reveal a gold watch-chain across his waistcoat.

I sit at my desk and start to write, then stop and glance at my grandfather, and then return to my writing. But something compels me to look up again, and catch his clear eyes looking straight at me: 'What are you doing writing on a fine day like this, boy?' he seems to be saying. 'You should be out in the fields.'

# Fat flitches, soft tempers

## February 1986

The pig and home-cured bacon used to be an important part of rural life. A couple of flitches of bacon, said William Cobbett around 1800, were worth fifty thousand Methodist sermons and religious tracts. They were a great blessing, he explained, softeners of the temper and promoters of domestic harmony.

In his day and afterwards many cottagers kept a pig for bacon. In the Forest of Dean they could feed on acorns and beech mast during the autumn, and sides of home-cured bacon helped several miners' families to withstand the hardship of the long strike of 1926.

Most farmers kept, killed and cured a bacon pig or two until comparatively recently. I suppose it was certainly the general practice thirty or forty years ago. The pigs were fed on barley meal and other plain wholesome food until they were large and fat, twenty score (400) pounds or more—twice the size of a modern bacon pig. Make him quite fat by all means; the last bushel is the most profitable, and lean bacon is the most wasteful thing any family can use, said Cobbett. Many a farmer and cottager followed his advice until well into the middle of this century, some believing that a pig was not fit for bacon until it could scarcely turn around in its sty.

Almost every parish had its pig-killer, perhaps the local butcher or a smallholder supplementing his winter income. The pigs were killed between October and March, preferably in cold weather, for there truly was a time when pigs bred for pork and bacon were slaughtered only when there was an 'R' in the month.

The killing of the pig was a gruesome affair. The animal knew its end was imminent, and with the approach of the pig-killer it started squealing piteously, a long tortured and tortuous noise, until it was silenced. Even after death it struggled on the bench while its cut throat spurted blood. Then the carcase was laid on the ground and covered with wheat straw which was set alight to burn its hair, after which began the arduous business of scrubbing.

A few days later the pig-killer returned to cut up the animal. Besides the flitches and hams there were six joints of pig-meat—it was always pig-meat with baconers, never pork. The head was used for brawn, the fry for faggots, and the chitterlings cleaned in salt and water and turned on a stick for several days. There was no waste. As they used to say, you could use every part of a pig except the squeal.

The household lived like lords after a pig-killing, but in those days with no refrigeration, neighbours used to swap joints, with meat given returned when neighbours killed their animals. And as Cobbett remarked, the squealing of pigs would usually be heard by others who might arrive in the hope of largesse.

Next came the curing, which involved quantities of salt. The bacon often had the merest vestige of lean, and was extremely salty when fried. Eaten at breakfast it would set up a man for his day's labours—but it would also give him a thirst for sharp home-produced cider. Boiled and eaten cold it sustained him during the day, or at the end of it.

Cobbett said that a man who could not live on solid fat bacon either needed the sweet sauce of labour or was fit only for the hospital. And so it was until forty years ago; countrymen lived, worked and thrived on fat bacon, beef and mutton.

Today such a diet is anathema to taste and medical opinion, and might indeed be thought to *fit* a man for hospital. Looking back, folks then did not suffer from seizures and coronary thrombosis in the way they do now, perhaps because their fatty food did not contain the residues of growth-promoting agents and all our other wonderful additives. For them the only additive was 'the sweet sauce of labour' and fresh air.

In short, the old-time countryman's food was wholesome. Recently a Member of Parliament complained that more was known about what went into a pair of socks than into food; compare this with the time when men who worked hard could and did live on a daily ration of fat bacon. It did not kill them—but without it their work might well have done.

# Hooks and sighs

## March 1986

Armed with a hedge-bill I started to cut back an overgrown hedge. There was a nip of frost in the air, the sky was clear and blue and a benign sun cast its warmth upon my back as I worked. It was the last Saturday in January, but it could have been a day early in March.

There was a time when I used to be deceived by such days into thinking spring was imminent. Spring is always sudden when it does arrive, but even in early March it is rarely imminent. With the impatience of youth I used to long for spring, but now I find the winter is gone too soon—gone before those winter jobs we meant to do could be finished, and often before they could be begun.

On this day stolen from March, however, at least I had started on one winter job. It needs a sharp eye, a sharp hedge-bill and swift upward strokes to leave a clean cut. If speed is the only criterion, then a machine is far better than my simple hedge-bill and muscle. But if a machine is cheaper than my simple hard labour, then how expensive my labour must be.

Machines have become a necessity; usually, these days, there simply is not the time or labour to do things by hand. But while we save time by using machines, we also lose the delight of using hand tools, as well as some pride in a job. If hedges cut by hand had been left in the torn and jagged state seen after the machines have been at work, the perpetrator would have been a subject of scorn and derision.

As I worked with my hedgebill I thought of the tools once in constant or seasonal use on almost every farm, but now rarely, if ever, used—hedge-bills and bill-hooks of so many types and patterns. Every district had its favourite, the Dorset bill-hook, the Gloucester,

Devon, Sussex, Suffolk, Kent or Ledbury. They came to fit their user's hand, and Richard Jefferies saw the bill-hook as almost an extension of the farmworker's arm.

There used to be a vast range of farming tools: spades for ordinary digging, for digging post-holes, several different types for ditching and tile-draining, open spades for cutting through heavy clay; tools for haymaking and harvesting; scythes and reaping-hooks—the tool most of us know as a sickle is invariably a reaping-hook, the sickle being smaller, with a serrated edge; and a variety of pitch-forks, some with three prongs and some with only two, always known to us as pikes.

The bow-saw and then the chain-saw replaced the large cross-cut saw, which needed two men to use it. The chain-saw is a fierce and noisy thing, but I never found any delight in using a cross-cut saw. Many trees would probably never have disappeared if the chain-saw had not been invented, for one man and a chain-saw can do a lot of damage in a hurry.

Machines save a lot of hard labour—in dung loading and spreading, for instance. Few farmers would willingly return to hand-milking, and if we were forced to I feel there would not be such talk of milk surpluses or quotas. Ironically, seeing large agricultural machines in the fields, many townsmen now concede that farmworkers *must* have some skill to operate them—yet the workers of old almost certainly had more real, varied skills.

As I grasped the handle of my hedge-bill, I thought of other handles that used to be held during the winter months—handles that required muscle but little skill, wooden handles worn thin and shiny through being turned round and round, the handles of chaff machines, root pulpers, cake crushers. Then there were iron handles of water pumps that went up and down, even starting handles of old tractors, which you sometimes turned and turned in vain until you were breathless on cold winter mornings. When I think of all these, perhaps, after all, I was a bit harsh about modern machines; it is just that I wish they were not so big and noisy, and could be sure that they are our servants and not our masters.

Postscript: a book about the Monmouth Rebellion published recently by Alan Sutton tells me that one of the rebels was named Humphrey Phelps. Perhaps he, too, was armed with a hedge-bill.

# That Man Again

## March 1978

In September, 1939, Bristol was a strange place—according to BBC producer Francis Worsley. That month, the corporation's variety, music, and Children's Hour production departments moved out to the city because of the war, presumably because it was believed to be safe.

Regional headquarters in Whiteladies Road were not large enough to house the influx, so the newcomers were scattered over Clifton, and every parish hall was wired and used for studios, staff and performers rushing from one to the other on newly-acquired bicycles.

*It's That Man Again* was first broadcast from Maida Vale in July, 1939, and three more productions followed without any great acclaim. It was the broadcasts from Bristol that turned *ITMA*—as it was then named, in deference to the wartime rash of initials—into a national institution. To those too young to remember, it is hard to explain the impact and popularity of the show; no television series has ever approached it. Almost everyone was an addict, including King George VI, who later invited *ITMA* to perform at Windsor Castle.

The first programme from Bristol went out on 19 September, after only two hours' rehearsal. At that time the scriptwriter, Ted Kavanagh, father of the poet P.J. Kavanagh, had not been passed by the Ministry of Information's security, and was almost smuggled in to write the show. It was broadcast live from Clifton Parish Hall, now long gone, and starred the soon-to-be legendary line-up of Tommy Handley, Jack Train, Maurice Denham, Vera Lennox, Sam Costa, and Jack Hylton's Band. The producer was Francis Worsley.

It was not long before Clifton, and the whole of Britain, were reeling under a mass of ringing telephones, slamming doors, catchphrases, and crazy repartee. While at Bristol, Handley was in charge of the mythical *Ministry of Aggravation and Mysteries*, and the *Office of Twerps*. Next door to the BBC in Bristol was a branch of the real Office of Works, some high-ranking inmates of which were not amused, especially when there was confusion with the mail.

Funf, the ineffectual spy, was featured in the second programme from Bristol. Funf telephones Handley: 'Funf,' he says, in a sinister, guttural tone, the result of Jack Train speaking into a glass. 'Funf?' asks Handley. 'Is that a name or a rude expression?' 'It is Funf, your favourite spy.' 'It may be Funf for you, but it's not Funf for me.'

Vera Lennox played Betty, the secretary. Years later, Mr Handley had another secretary, the formidable Miss Hotchkiss. A real Miss Hotchkiss, also a secretary, who lived near Painswick, wrote in to say that at last she had some standing in her office.

Maurice Denham was Mrs Tickle the charwoman, and Vodkin the inventor. Sam Costa was Lemuel the office boy. And Jack Train was Farmer Jollop, and Fusspot the civil servant, as well as Funf. *ITMA* had left Bristol before the advent of Mrs Mopp ('Can I do you now, Sir?'); Colonel Chinstrap ('I don't mind if I do') and Ali Oop, with his naughty postcards ('You buy nice postcard? Very chummy, oh lumme!').

Most of the material was topical. Jokes were inserted at the last moment, and everything was performed at speed, as it had to be; there were some thirty-five weekly shows, each of half an hour, through each of the war years. Much of it would be incomprehensible to us now, and I dare say we would find the jokes very corny. But not so back then. To a blacked-out Britain, *ITMA* was a beacon—the brightest spot in six grim years.

The catchphrases were infectious, and repeated everywhere, endlessly. To use them was the height of drollery, or so many supposed, through convulsions of delight at their own apparent wit.

The first series from Bristol ended in February, 1940, by sending the Ministry of Aggravation and Mysteries into the country, pursued by the sinister Funf. Soon afterwards the period now known as the Phoney War—but more commonly the Bore War—came to an end.

Hitler swept across Europe, and Britain found herself in real danger of invasion. The male members of *ITMA* joined Bristol's Local Defence Volunteers, later the Home Guard—which prompts the thought that perhaps *Dad's Army* is TV's nearest approach to *ITMA* in terms of popularity. It is interesting to note, too, that *Dad's Army* was reputed to be the Queen's favourite programme, just as her father admired *ITMA*.

Before 1940 was out, though, heavy bombing prompted the BBC hierarchy to think again about Bristol as a rural refuge, and with broadcasting almost impossible, it was at last decided that the show should follow its mythical Ministry into even deeper country. One day in February, 1941, a special train left Temple Meads; *ITMA* and the whole variety department were *en route* for Bangor, North Wales.

# The raucous sound of reassurance ...

March 1980

The rooks came to live near us about twenty years ago. They moved into the big old elm tree at the back of the house. Unlike most elms this one had a spreading habit; from a distance it looked like a very tall oak tree.

Within a few years there were sixteen nests in the tree. Rooks, it is said, start rebuilding on the first day of March, except when it falls on a Sunday. I cannot recollect our rooks being Sunday Observance conscious, but they did begin work on that date, give or take a day or two.

If March came in like a lion they had a rough time of it, blown off course while carrying twigs. They'd caw raucously—which I took to be in protest over the weather—or glide with outstretched wings. If the month came in like a lamb they didn't make much better progress. Most birds work hard, rushing to and fro to complete their nests, but not rooks; there's something indolent about the building of a rookery. They spend a lot of time gossiping, idly watching the day go by, or another rook fetching a twig. Some weren't above fetching twigs from a neighbour's nest, either. Like W.H. Davies, they seemed to ask:

> What is this life if, full of care,
> We have no time to stand and stare.

We became fond of our rooks; their cawing may sound harsh, but how evocative it is. Though so different from the soft, seductive cooing of wood pigeons at fall of dusk, it can cast much the same spell; reassuring, timeless, strangely comforting. And what a hubbub the young rooks made in the evenings.

Then the Dutch elm disease came; our old tree survived longer than most, but at last it succumbed. Most of the branches were dead, but it still managed to put forth leaves on a couple, and the rooks still nested there. The following year it was completely dead, and the rooks tried building in the neighbouring oak—without success. The nests kept falling apart and tumbling down, and the rooks went away; I don't know where they went to or what they did; perhaps they didn't breed that year. The following year some returned and built in the oak tree or the willows, and with good success. Obviously they'd learned another technique to deal with the different branch formations, necessity being the mother of invention.

What has happened to the English apple? England has the best climate in the world for apple-growing, and there was a time when the flavour of ours surpassed all others. But does it now—and, if it did, would people still buy those Golden Delicious apples from France? Perhaps they would, but who'd have thought the public would have grown so fond of a green dessert apple. Someone must have had a quirky sense of humour when they named that apple Delicious, too—certainly if they'd ever tasted a Cox Orange Pippin, a Blenheim Orange or a Ribston Pippin.

Nowadays, most Cox have lost that distinctive flavour and their looks. Is it just my fancy, a kind of nostalgia for lost youth, when any apple tasted good? I don't think so; pick a ripe Cox from a neglected old tree, and you'll find the old flavour. The apple will have more colour than the modern Cox, too—talking of which, is there a more handsome apple than Charles Ross?

Modern Cox? Surely a Cox is a Cox is a Cox. Yes, but different rootstocks exert different influences on apples, I'm told. Today's Cox tends to be bigger, not so colourful and more bland.

Another reason for the loss of flavour is because commercially grown apples are picked far too early. You'll see Cox for sale almost as soon as September is in, and they're a disappointment. The moral is: grow your own and let them ripen before picking.

A third explanation suggested by a man with a lot of experience in fruit growing is that the soil can supply only so much flavour, and with successive heavy crops that flavour is diluted in each individual apple. I must admit I found this unconvincing at first, but there may be something in the theory. Excessive use of artificial nitrogenous fertilisers on grassland may produce a lot of grass, but it does not necessarily produce a corresponding increase in milk or meat, the real product of grassland. Doubling the yield of wheat means twice the bulk, but the nutritional value, hundredweight for hundredweight, may be less. It could be the same with flavour.

# My golden memories of Merrylegs and Huffcap

## March 1982

We have a four-acre field named Love Field; why Love Field I do not know, and do not really want to know, because the reason may be less interesting than my speculations.

In spite of its romantic and intriguing name, the field is, and in living memory always has been, called the Barlands, and this time the reason is obvious. Several Barland pear trees once stood in it, and there are still a few, three times the height of telegraph poles and probably some three hundred years old. In the spring their blossom makes the most attractive sight and gives off a pleasant aroma; most years they produce a crop of small, hard, round pears.

The Barland is a perry pear. Three hundred years ago it was well known and widely planted, and its perry was of high repute. The Barland was named in *The Whole Art of Husbandry*, published in 1708, as one of the five varieties of perry pears esteemed above all others for its vinous juice, and William Marshall, in his *Rural Economy of Gloucestershire*, published in 1789, wrote; 'The Barland is in great repute as producing a perry, which is esteemed singularly beneficial in nephritic complaints.' Thomas Rudge's *Agriculture of the County of Gloucestershire* (1813) agreed: 'The liquor has great strength, and a peculiar flavour, grateful to the palate.' He, too, spoke of its medicinal qualities, especially in cases of gavel.

Perry pears used to be grown almost in every parish of Gloucestershire except the Cotswolds, and perry-making was an important part of the county's rural economy, particularly in the seventeenth and eighteenth centuries. Perry and cider were sent to London and Bristol, from whence to be shipped in bottles to the East and West Indies and other foreign markets.

Marshall mentions a particular tree which one year produced three hogsheads of perry. He also gave the opinion that low wages and large amounts of cider and perry were the curse of Gloucestershire's agriculture: 'A Severn man's stomach holds exactly two gallons, three pints.'

Common perry in Marshall's time was sold for as little as a guinea for three hogsheads, or less than a penny a gallon, a hogshead being 110 gallons. Squash-pear perry was usually £5.10s a hogshead, often a shilling a gallon and sometimes twice that amount. The best squash-pear perry was practically indistinguishable from champagne, the produce of such as the Arlingham Squash pear, the Staunton Squash, and most famous of all, the Taynton Squash.

Another local perry pear still well known today is the Blakeney Red, also known as Red Pear in Blakeney itself; Painted Lady, Painted Pear and Circus Pear. This last name is an indication and a warning of its effects: like a circus horse, it went once round and out, especially if the pears came from the rich riverside land. From the poorer Forest soil the perry was more stable, even if my pun is deplorable. The Blakeney pear also cooks well, and has been used for canning and pickling. During the First World War it was used as a dye for the manufacture of Khaki.

There are pears carrying the names of Gloucestershire villages, or of drinks: Brandy, Claret, Gin, Port, and Sack. Some, like the Blakeney, have several names, varying from district to district, and many are descriptive, outlandish or far more intriguing and suggestive than that of our four-acre field.

Blacksmith; Bloody Bastard; Brown Bess; Clipper Dick; Dead Boy; Ducksbarn; Grandfather Tum; Green Horse; Huffcap; Pig; Snake Pole; Sow; Treacle; Holmer or Startle Cock; Rock or Mad Pear; Lumber or Steelyer Balls; Rumblers or Rumble Jumble; Nailer; Stinking Bishop; Tumper. The best and happiest of all is Merrylegs. 'Of little character,' the Long Ashton Cider Institute said of it in its *Perry Pears* (1963), but with a name like that it must have had some character, even if only a bad one. It must have been the cause of some rollicking characters, too.

Long Ashton's comprehensive book, as far as I am aware, is the only one to deal with perry pears; indeed, the research, advice and help of Long Ashton has been of great value to the farm cider and perry industry generally. Financed almost entirely by the industry, it is the only body in this country to carry out any objective scientific work on cider and perry, research for which it has built up an international reputation. Now the Agricultural Research Council has plans which would mean the closure of its cider and perry section, and it is only to be hoped that the opposition of independent West Country cider-makers will prompt second and better thoughts.

# Jefferies: misunderstood,
# but not unappreciated

## March 1981

'It was bitterly cold and even snowed a little. I wrapped my rug round my knees and wrote. I got into bed just as the Abbey bells chimed 5 o'clock in the morning. Strange musical old bells they are, very soft-sounding and melancholy, as if telling a requiem over the old monks who built the abbey. It is a rare church—almost a cathedral.'

The writer was Richard Jefferies, and he was twenty-one years old on that day he heard the bells of Cirencester. Twenty-one and with less than another twenty years to live, but during those years he was to become the most versatile of country writers. And like most versatile writers, his reputation has suffered because of it.

Some, Gilbert White for example, may have excelled him as naturalists, but no one has provided so sweeping, detailed and intimate an account of all aspects of the country; of farming, farmers and farmorkers, landowners, gamekeepers, poachers; of almost everything that grew, walked, crawled, fled or swam.

Perhaps, too, no country writer has been so misunderstood. Nostalgic he was, but he did as much looking forward as backward. And while he has been branded sentimental by writers far more sentimental than he, there is nothing sentimental, to give but one example, about 'One of the New Voters' in *The Open Air*.

It was *The Story of my Heart* that turned so many readers away from Richard Jefferies. It is not an easy book, too lush, too confused, too mystical for most people. But in it he tried to explain the unexplainable, and he should not be condemned because he did not succeed entirely in that.

Richard Jefferies was born at Coate, Wiltshire, in 1848, the son of a not-very-successful small farmer. But although he was to write so knowledgeably and lovingly about the country, he had no inclination to work on the land; in any case, his indifferent health would have precluded it.

Having done some reporting for the *Wiltshire and Gloucestershire Standard*, he became a member of the staff in 1868, which probably accounts for his being in Cirencester on his twenty-first birthday. He stayed with the paper for two years and rejoined it for a short period later; he also wrote a history of Cirencester which appeared in serial form in the paper.

*Hodge and his Masters*, which also first appeared in serial form, was published as a book in 1880 and has recently been reprinted; described as a classic of English farming, it is almost his best book.

The market town, Fleeceborough, is Cirencester, undisguised except in name: 'The place is a little market town, the total of whose population in the census records sounds absurdly small; yet it is a complete world in itself ... Enter Fleeceborough by whichever route you will, the first object that fixes the attention is an immensely high and endless wall ...

'From time immemorial both Hodge and his immediate employers have looked towards Fleeceborough as their capital ...

'All the produce—wheat, barley, oats, hay, cattle and sheep—is sent into the capital to the various markets held there. The very ideas held in the villages by the inhabitants come from Fleeceborough; the local newspapers published there are sold all round ... The farmers look to Fleeceborough just as much or more ...

'The antiquities of the old, old town are kept for it, and are not permitted to decay ... the Roman villas ... all the fragments of past ages ...

'There was a famous poet (Pope) who sang in the woods about the park; his hermitage remains ...'

*Hodge* was written a hundred years ago, and country life and farming have changed; changed they had when Jefferies wrote the book, and changed, indeed, they have since. Yet reading passages of the book, except for a word or a phrase here and there, the sentiments could have been expressed today.

Jefferies writes of the new system of selling milk, instead of making butter and cheese; of efforts being made to extend the area available for feeding by grubbing hedges; of drainage and fresh grasses sown, and hay of better quality.

Cowyards have been improved—and so have the cows, which yield large quantities of milk, not to mention the machines which mean that the most can be made of good weather; the hay is gathered in quickly, instead of lying about until the rain returns.

New roads have been laid across farms; dairy farms have been swept and garnished, and even something like science has been introduced to them. Artificial manures are spread abroad on the pastures, and the dairy and arable farmers are bringing modern appliances to bear upon their business.

I said that *Hodge and his Masters* was almost his best book; it would, I think, have been his best without doubt had he not favoured the masters at the expense of Hodge. Jefferies was guilty of failing to tell the whole truth, and Hodge has not been treated fairly, but apart from that it is a good book with a wonderful cast of country characters and closely observed details of the rural scene a hundred years ago.

*Hodge and his Masters*, *The Hills and the Vale*, *Wild Life in a Southern County* and others are available in paperback after being out of print for many years. Richard Jefferies died in 1887 at the age of thirty-nine, poor and ill for most of his life and often misunderstood; now, at last, it seems as if his worth is being appreciated by new generations.

# Doing and chewing

## March 1979

We were standing in a large, lofty old timbered building. My friend was showing me a bunch of his fattening bullocks; all sleek, rotund and placid, knee-deep in straw. 'They're doing,' he said.

This remark may have puzzled a townsman, to whose eyes the bullocks would not have been doing anything in particular, except chewing their cuds. But to farmers, 'doing' means thriving, an expressive term and one we use with relish, as we relished the sight of these animals then.

A bunch of bullocks doing well fills and warms the heart of a farmer. Only a good, level crop of ripening corn with the heat shimmering over it gives greater pleasure. We watched them with silent satisfaction, finding comfort of our own in their well-being.

'It's a funny thing,' my friend said, 'but cattle always do well in this old shed, better than in our modern ones. They always seem to sweat in those, even though they are often cold. It's warm and airy here, and sheds like this fit into the landscape, like the old houses; they were built in local materials.

'This old shed is roomy—you can get a tractor in here. That's the chief fault with some old buildings; you can't use a tractor to clean them out. We've used this place for all manner of things; bullocks, pigs, we've stacked sheaves of corn in here, and hay and straw, and stored loose corn, potatoes, and roots.

'That's the best part of it, it's suitable for so many uses. I think it's a mistake to spend a lot of money putting up a building that can only be used for one job, or is incapable of being converted. You never know; times and farming methods change.'

I thought of my own farm buildings. The stable is now a milking parlour and dairy, the old milking shed has been turned into loose boxes. A wagon house is occupied by Gloucester Spot sows. My big old brick barn, after a variety of uses, has reverted to its original function of storing corn, and it is there, too, that we grind, mix and store our cattle food.

We took our leave of the bullocks. As he closed the heavy wooden door of his useful old shed my friend chuckled and said: 'Eccentric landowners with more money than sense used to build follies. Today it's farmers who build 'em; but they call them silage towers.'

# Salutation

March 1978

'Why do you keep putting your hand up, as if you are saluting?' a friend asked, while we were walking round the farm. 'Oh, but I *am* saluting.'

He looked puzzled; no, not so much puzzled, more as if thinking: 'He's gone; but then, he always *has* been a bit eccentric.' So I hastened to explain that I was saluting magpies, in order to avert misfortune. Of course, I'm not superstitious, it's just that I don't believe in tempting providence. And in these cases it doesn't do to be shy, like the man being taken to catch a train who didn't salute a magpie because he thought his companion would laugh. He missed the train and his companion said: 'There, I knew we should have saluted that magpie.'

I have heard that an onion is an effective safeguard, but it could be inconvenient, always having an onion in your pocket. It does mean a lot of saluting, though; we've magpies everywhere, clutches of them, flying about and strutting like fat, pompous aldermen. A few would be acceptable, but such numbers must reduce the population of smaller birds; magpies, after all, prey on their eggs.

We have far too many jackdaws, too, and they steal the food from the hens and pigs, as well as launching into a crop of barley, at the milky stage, just like a combine. Such cheeky birds; they never fly right away when shouted at, but merely perch at a safe distance and laugh, and jeer, and swear.

A long time ago, we made a cage to catch them, and put a live jackdaw in it as a decoy. Jackdaws came and peered down the funnelled opening, but didn't go inside. More jackdaws paced round the cage, and then flew away to return with more jackdaws. They inspected the cage, and made angry noises. Still more arrived; soon the air was black with jackdaws.

They held a mass-meeting in the adjoining field. They were angry; they were noisy; they hovered in the air; some prowled round the cage; they held another convention. Not for anything would I have ventured into those fields. That crowd, now many hundreds strong, was in an ugly mood.

Next morning there was one jackdaw—the decoy—in the cage. It was dead. We never tried to trap jackdaws again.

# Golden stars of springtime

March 1984

Aconites, periwinkles, snowdrops; all these are signs of the return of spring, but to me the real harbinger is the celandine. It makes its first shy appearance in February, but does not reach full glory until March, when it spangles the banks of hedgerow and wayside with brave golden stars.

March can be as much winter as any month, but the celandine, undeterred by cold, wind or rain, proclaims the promise of spring, no matter how long-delayed its arrival may be. A little later, anemones, more happily called wind flowers, wild daffodils, and the green in the wintered hedge signal its imminent coming more accurately than any calendar date.

Lady Day, March 25, is a date often ignored by modern calendars, but once it was an important one. Michaelmas Day, September 29, or Lady Day were the times when farm tenancies began, and when farms exchanged hands. In our time Michaelmas was probably the most popular, but in their surveys of agriculture in Gloucestershire both William Marshall (1789) and Thomas Rudge (1813) give Lady Day as the more popular date.

Marshall details the terms of tenancy in the Vale; restrictions on selling straw off the farm; all muck, dung or compost to be used on it; all the corn and hay grown on it to be ricked and housed on the premises; no pasture to be ploughed; all hedgerow trees to be preserved; a certain number of willow trees to be planted each year, and so on. On the Cotswolds, tenants were permitted to sell wheat straw.

In Marshall's time much of the Vale was enclosed, so there were comparatively few landlords. Soon, however, there was a large-scale enclosure movement in England, which meant the virtual destruction of the peasant class. It was claimed that enclosure of the commons brought a better system of husbandry, and no doubt it did. It also created a class of landless labourers, hardship and poverty, and made the countryman dependent on money.

Arthur Young (1741–1820), the first secretary to the newly-created Board of Agriculture, was a fervent advocate of enclosure. Later he had regrets. 'The poor look at facts, not meanings,' he wrote, 'and the fact is that by nineteen enclosure Bills in twenty they are injured ... while the father of the family is forced to sell his land.' Even if they were diligent they were given no leave to build a cottage, no land for a cow, not half-an-acre for potatoes, said Young. The dispossessed could look forward to nothing but the parish officer and the workhouse.

Small squires and farmers also lost their land through the enclosure movement. Most of the fields were enclosed by the end of the nineteenth century, and by 1910, some 90 per cent of land in England and Wales was cultivated by tenant farmers. Ironically, it was only a few more years before the landlord-tenant system began to crumble.

The landlords, no doubt, were autocrats, though many were possibly benevolent ones; Coke of Norfolk was the outstanding example, and there were others. The tenancy agreements were restrictive, often designed to protect the sporting interests of landlords, but they also protected the land itself; and whatever the defects of the system, it did give thought to the future as well as the present. It did, too, create the countryside which now stands in danger of destruction.

Under the old system land was a commodity to be protected; under the present, it is something to be exploited. The first philosophy thought of tomorrow, the present one thinks only of today.

In spite of the many defects of the old landlord-tenant system, then, and the manner in which it was often implemented, I have no doubts about which was the better. Ideally, though, I would go back even further in time, and settle for the days when we were still a nation of peasant farmers.

# The Gloucestershire poet

March 1983

There will be a notable Gloucestershire centenary five years hence, for it was on March 26, 1888, that Frederick William Harvey was born at Hartpury. The family moved to Minsterworth in 1890, the year in which his friend Ivor Gurney was born in Gloucester, and the early careers of both became intertwined when they struck up a friendship at King's School, Gloucester.

Later Harvey went to Rossall and progressed to study law, while Gurney opted for music. There was divergence in their later lives, too, with Gurney dying in London after a long mental and physical illness in 1937, and Harvey, an established Forest personality, surviving him by twenty years.

There were parallels, true in that both won considerable acclaim as poets in their youth, were overlooked in later years, and by the early 1980s had earned some posthumous recognition, symbolised by the fact that plaques had been erected to both of them in Gloucester Cathedral.

Last year, with the publication of his collected poems by P.J. Kavanagh, the poor tormented Gurney at last attained his deserved position in English literature. Harvey, I feel, has yet to be accorded the honour he genuinely merits.

F.W. Harvey's first volume of verse, *A Gloucestershire Lad*, was published in September, 1916. Much of it was written at the front line, but as his commanding officer said in the preface: 'Mud, blood and khaki are rather conspicuously absent ... What he does think of is his home':

I'm homesick for my hills again—
  My hills again!
To see above the Severn plain ...

Ivor Gurney, also in the front line, set the poem to music, and when
Harvey was posted missing, Gurney wrote:

He's gone, and all our plans
  Are useless indeed ...

John Lehmann describes the end of this poem as 'almost unbearably
painful.'

Harvey was very much alive behind enemy lines, and his
next volume, *Gloucestershire Friends* (1917), came from a German
prison camp. Included in it was that courageous poem *Solitary
Confinement*:

No mortal comes to visit me today,
Only the gay and early-rising sun.

*Ducks and Other Verses* appeared in 1919. *Ducks*, of course, is his most
famous poem, but I cannot help wonder if its popularity did not do
him as much harm as good, by obscuring other and more thoughtful
works. *Ducks* is good and whimsical, but by being labelled as its
author, perhaps Harvey also runs the risk of being dismissed as
such.

In 1920 Harvey published a book of prose, *Comrades in Captivity*,
a record of life in seven German prison camps. The timing was
perhaps unfortunate, in that the war was still too near, books about
it not becoming as fashionable until later in the decade. Another
book of verse, *Farewell*, appeared in the following year, and his last
important volume, *September and Other Poems*, was published in
1925:

O Lord, within my heart forever
Set this sweet shape of land and winding river ...

This and *November, On Painswick Beacon* and others should have
secured him a lasting reputation.

*September* was his last volume from Sidgwick & Jackson, who published many of the war poets. But a small collection, *In Pillowell Woods*, was published in 1926 by Frank H. Harris of Lydney, by which time the poet was practising law in that town and living at Pillowell. Later he moved to Yorkley.

He continued to write poetry, some of which appeared in magazines and newspapers; a selection of mostly published work, *Gloucestershire*, appeared in 1947, and *Forest Offering*, containing seventeen unpublished poems, came a few years after his death.

Today it is not so much that Harvey is under-rated; he is no longer rated at all. Fashion can be blamed, but only partly, for he sang of things that can never be out of fashion so long as life, and love, and laughter live.

# Progress on the rampage

April 1984

There are certain phrases and words which, with constant use, become sacrosanct—and eventually meaningless, too. Efficient is one. Everything must be efficient, and as often as not, to be efficient, it has to be big, too. But efficient for what? And what is efficient?

Modernisation is supposedly efficient. But a machine which needs only one operator and will do the work of fifty men—is *that* efficient in the eyes of the forty-nine who are put out of work? It can be argued that it cheapens cost, but does it do so in human terms, and does it produce better work?

Computers, I am told constantly, are wonderful things. But I am also told constantly that errors which rarely happened before are 'because of the computer.' I cannot excuse my errors so easily, so perhaps I should get one. Then I could make mistakes galore and everyone—except for old-fashioned people like myself—would condone them.

I am told constantly, too, that you cannot stop progress. Progress is another sacred word, a concept not to be criticised. Like efficiency, it is good—but again, what is progress? There was a time when progress meant improvement, but now there is a world of difference between the two. Progress all too often means a lowering of standards of life and contentment. As Orwell once observed, progress is a menace to everything—a process in which everyone goes faster and faster towards they know not where.

I have seen both progress and improvement come to farms and villages in Gloucestershire. For a time they were almost the same, or at least they were abreast of each other, but now we are in danger of losing as much as we have gained, or more.

Progress and large-scale mechanisation, aided and encouraged by grants, have transformed farms into factories and turned men off the land. Several small farms have become one large one. Straw is burned, instead of being returned to the land, where it belongs. Animals have become units for exploitation. Farmers look enviously for more land; and rent money in order to buy it at a price beyond what it can return in terms of real husbandry. Land, the most precious of commodities, has been reduced to stocks and shares.

Old skills have been lost. Machines, wires and pipes have entered fields and houses. Then come the replacements, bills, water rates, licences, forms and officials, all of which and all of whom cost money. The resultant standard of living is better, but as progress increases, so do the pace and prices, and we are all on the treadmill, working harder, or casting around for work with growing desperation.

Without progress we cannot have more and better amenities, so we should not grumble. But we do, as we see our old amenities vanishing; public transport, schools, post offices, shops, and many other faces of the rural way of life. There are, I admit, some gains, though thay escape me at the moment; but the losses are great—and among them is that old-fashioned commodity, content.

Instead we have efficiency and progress—all right, I suppose, so long as they do not gallop out of hand and cause damage. Runaway horses were never any use to me; I much prefer that trusty old workhorse, improvement.

# Bewitched

April 1978

One of the frustrations of writing fiction is the fact that many of the finest incidents from real life cannot be used—because of the law of libel, or the danger of hurting someone who does not deserve it, perhaps, or more often, because they are simply not believable. Fiction, unlike real life, must be believable.

Take the case, some years ago, of my cow Buttercup. Buttercup was a placid, good-natured cow; nothing ever upset her until that one afternoon. She came in to be milked as usual, went to her place in the shed and waited patiently for me to fasten the chain round her neck. Within ten minutes she was in a wild frenzy, tugging at her chain, bellowing frantically, soaked with sweat and shaking with fear. We knew of nothing that could have caused such alarm—but there was no doubt about it, she was quite simply terrified.

We could not leave her in the stall like that; we had to release her. She stumbled out into the yard, still bellowing and shaking and picking her feet up in a peculiar manner, as if the ground beneath were electrified. We managed to coax her into a loose box, well away from the cow shed, and eventually she lay down, in a cowering heap.

I telephoned the vet. When he arrived, about an hour later, she was much quieter, though still trembling and coated in sweat. He examined her, but could find nothing organically wrong. 'I don't know what the devil's the matter with her,' he said. 'It's obvious that she's been frightened almost to death. You'd think she'd been bewitched.'

Now, what he did not know, and for that matter, what I had forgotten until then, was an incident of that previous morning.

There had been rain, and the roadmen who had been working close by had come into my barn for shelter, as they had done on similar occasions. And as it was wet, I had stood in the barn talking to them. One of them, a comparative newcomer to the gang, a tall, gaunt man with deep-set, far-away eyes, had said, apropos of nothing, that he could foretell the future, put spells on dogs and perform similar strange and occult deeds.

I was sceptical; indeed I scoffed at him. The other men remained silent, and at a much later date two of them told me it had been unwise to make light of their companion's powers; once he had foreseen a man's death in a ditch, and within a week his body had been found, not in, but close enough to a ditch to make people remember his words with apprehension. This strange roadman had obviously been put out by my doubts, for he did not speak to me again. The sun came out, the man left, and I returned to my work and forgot about the incident ... until the vet's remark.

And there you have the bare and true facts about my cow Buttercup. What made the vet say what he did? *Did* the roadman bewitch her? And lastly, how could I possibly expect anyone to believe *that* story, if I used it in a book?

# A Cotswold diary

In the early 'thirties C. Henry Warren gave up his job in London and went to live in a cottage at Stockend, on the Cotswolds. He was the cause of much speculation in that isolated community; his neighbours wondered what he did for a living. Four years later, some of them received copies of his book *A Cotswold Year*, which H.J. Massingham saw as 'the best book upon our revered Cotswolds.' Since then there have been many more books about the Cotswolds—one could say that of the making of books about the Cotswolds there is no end—and yet none, in my opinion, has surpassed Warren's.

Warren's is in diary form; the reader is taken on a tour of the Cotswolds, but not the usual, rather dull tour of the guide book. Like the view from his cottage window, his book is not confined to the hills, but the best of it is about the people in and around Woodend, his name for Stockend. Farmer Flack, scratching a living, but quite content on his few stony acres. Jesse Gable, repairing dry-stone walls, busy in his garden or making a concrete barn owl. 'He weighs nigh on a hundredweight,' said Jesse. 'I shall fix'n on the wall by the front door.' Mr Dane, who kept the village 'general'. 'You buy more than mere groceries, or whatever it is, when you buy at his little shop,' wrote Mr Warren. Those who are lucky enough to have dealt with such stores will recognise a complete description of them in this one sentence.

I met Jesse Gable—and his owl—three or four years after the war. He was busy oiling the hinges of his garden gate as he told me: 'Stockend have looked up since Mr Warren was here. It was a poor old place then.' Henry Warren was then living at Finchingfield,

Essex, and had written several more books, the most notable, perhaps, being, *England is a Village*; several of his works include a visit to Gloucestershire. After a long and painful illness, which caused him to lose his voice, he died in 1967, just before his last was published.

The passing years have been rough on the Flacks, the Gables, the Danes and their kind: bullied by successive Governments, harassed by bureaucracy, hounded by petty officialdom and squeezed by big business. Some survive, and now that the crazy system that dealt them such blows is tottering they may still emerge triumphant.

# Life and lanes need corners

Last April's atrocious weather—those blizzards and all that snow at the end of the month—shook my faith. Until then I had always had some trust in the weather, believing that there was a kind of consistency even in its inconsistency.

Neither my cows nor I have yet recovered properly, nor the pastures which were turned to mud during May. Only a decent spring this year will set things to rights; with luck the cows will be out at grass this month, and there will be no need to hurry out to bring them back indoors.

During March, when the stocks of hay, silage and straw have dwindled, how eagerly we look forward to the day when the cows can be turned out to grass—and how eagerly the cows look forward to that day. There comes a time when they smell the scent of spring, of grass and of freedom, and they become restless in the confines of the yards.

Yet as that day becomes imminent I view it with a certain reluctance; one gets into a routine, a winter rut and whatever the reason—increasing age or indolence, I suspect—I find I am loth to leave it. I would not have been bold enough to admit this, however, had not a neighbour said he felt the same way.

Once the cows are at grass, sooner or later they have to go on the road to reach the fields. Now it is an odd thing, but the cows have only to step on to the road and cars appear from both directions, especially on Sunday afternoons. The majority of motorists are patient enough—and patience with cows is always the best and quickest tactic.

There are some drivers who do not realise this; neither, I think, do they realise that anyone else, least of all cows, has any right to be on the roads. To them, all I can say is that if they do not like being hindered by a herd of cows going about their business, then they would do better to avoid country lanes.

There is something that creeps into some people, once they are behind steering wheels, that makes them impatient, rude and inconsiderate. Often, I have noticed, the slower a person is normally, the faster he or she will drive a motor car. They will not understand that country lanes are not meant for speed, that something may just be around the corner. Have you ever noticed their look of astonishment—even indignation—when suddenly confronted with something round a corner?

It is no longer safe for children to ride bicycles on country lanes—and remembering the pleasure I had in this when a boy, I am angry that it is denied present-day youngsters. It is scandalous; we should do something about it—though just what, I do not know. If there were a speed limit of 30 m.p.h., how could it be enforced? Commonsense dictates such a limit, but the craze for speed does not listen to commonsense.

Cut down the hedges, widen and straighten the roads is one suggestion, but this is no remedy at all. Quite the reverse; it will only increase speeds when the object should be to decrease them, for speed has become an obsession. The faster people go, the less time they seem to have, and the more ill-tempered they become. They drive recklessly past here with a scowl on their faces; but last January, when the snow kept their vehicles in the garage, those same folk walked with a smile and had time to stop and speak.

The hedges, the narrowness and the corners are all part of the charm of country roads. They are not dangerous; it is those who complain who are dangerous. To travel joyfully should be the maxim, and if you travel slowly along these roads it is possible to travel joyfully, enjoying the flowers of hedgerow and verge, all springing up again since the council has stopped spraying and cannot afford to keep trimming the verges.

How dull our lanes and our lives would be without corners, without the thrill and delight of whatever lies around the bend—and if it proves less than a delight, then it is all the better for having been hidden for a while. A garden is perhaps the best example of the charm of corners. The art of planning one is not to reveal everything at a glance, but to have curved borders, corners with mystery and magic. Half the pleasures in life are in anticipation; banish corners and you banish half life's pleasures.

Banish corner, too, and you banish the charm of the country road—and if a herd of cows happens to be making its slow, sinuous way along it, be patient and think of these words of Richard Jefferies:

There is the grass, and the wheat, the clouds,
the delicious sky, and the wind, and the sunlight
which falls on the heart like a song.

A few minutes lost can be a few minutes gained.

# A hard un this time

April 1981

This year January stole days from March and even April. The radiant sun, the mild weather and the song of birds led one to expect the hedgerows to burst into green leaf, and the rooks to start inspecting their nests.

'I like winter in winter,' remarked the little man in the pub on market day. 'I like winter in winter, not in April.'

'I thought it would be a hard un this time,' said a farmer.

'Time enough yet,' someone else muttered. 'The hard un in forty-seven didn't start until the twenty-third of January.'

'I've been an old countryman all my life,' said the little man, 'and I like winter in winter. When we get spring in January we're likely to get winter in April.'

'I was laid up for a week that year,' said a large grizzled man quietly.

'You know where you are if you have winter in winter an' spring in April an' I'm a man who likes to know where I am,' the little man continued, raising his voice.

'You're in the pub, Alf,' said a tall, thin man.

'I know I'm in the pub, Bert.'

'You usually are, if not in this un, 'tis another, so I can't see what you're worrying about.'

'And I had a milk round then, we used to dip the milk out of the churn in those days,' the large grizzled man said. 'An' my boy had to deliver the milk. An' one woman told him to put three pints in her jug an' he told her he couldn't 'cos her jug was a quart size. And she told him as he was a great soft boy and 'course he could 'cos his father always did.'

'It's all the same today,' said the little man ferociously. 'You don't know where you are.'

'I'd been doin' it for years,' said the large grizzled man. 'Milk was rationed during the war an' 'er always wanted more'n her share an' though there's them as may say it wasn't right, her was satisfied an' so was I, an' if we was both satisfied I can't see as there was much harm in it.'

''Ectares an' litres an' metres,' snorted the little man. 'I don't like 'em, you don't know where you are with 'em, 'tis all done to confuse us.'

'I don't like 'em,' said the farmer.

'None on us like 'em,' snapped the little man.

At this point the little man grew very excited: ''Ectares, litres, metres, VAT, bah! Men drinking lager. Scampi, chicken in the basket, pah! An' rules an' regulations an' laws an' more laws! I been an old countryman all my life an' I don't like it. What be they doin' to us I want to know.'

'If I had my way,' said a man who had not spoken before, 'I'd put all of 'em, the politicians an' experts an' them as keep havin' conferences an' all of 'em as keeps on beggarin' us about, I'd give 'em all a spade an' put 'em ditchin'.'

'They 'ouldn't know 'ow to,' said the little man.

'Then they'd have to learn just as I've had to learn their old nonsense, except t'wouldn't be nonsense as they was learnin' an' after a few weeks of ditchin' they'd have had a chance to think things over an' be more sensible fellows an' the country 'ould be the better for't to my way of thinkin'.'

The little man turned to me and demanded. 'Do you believe in huntin'?' Without waiting for an answer he adopted a posture as though aiming a gun and asked 'Do you believe in shootin'?'

'Well ...' I began, but before I could go on he rapped: 'Football? I don't believe you believe in huntin', shootin', or football.'

'Croquet is the only sport I've ever ...'

'Croquet? Croquet? What kind of talk is that? Croquet!' The little man spat the words out scornfully.

A tap on my shoulder and I turned to find an elderly farmer smiling at me. 'I quite agree with you,' he said. 'It's a delightful game. And the ladies. Tea on the lawn.'

'Cucumber sandwiches,' I added.

'And a drop of something afterwards,' said my new friend quietly, and smiled.

# Wealth accumulates and men decay …

## April 1986

A couple of months ago the retiring president of the National Farmers' Union spoke about the state of agriculture in this country—and threw in a scalding indictment of the present Government's farming policy.

Farmers, he said, are faced with uncertainty overlaid with confusion. He was anxious that Governments should not be allowed to throw agriculture on the scrapheap, and concerned that EEC proposals discriminate unfairly against farmers in Britain. He also warned that people in the countryside, as much as in inner cities, can and do suffer from recessions.

No doubt it was a good speech—but it would have been a better one had it been made years ago. A cynic would say, with some justification, that it was made now only because the big farmers are about to feel a chill wind—one which has been felt by the smaller farmers for quite some time.

The big boys are being faced with uncertainty. Many small farmers have been faced with much worse, and thousands have already been thrown on the scrapheap. I do not recall the president making an impassioned speech on their behalf.

This is no criticism of those NFU county branches who have tried to warn of the danger threatening the small farm—the small farm which is the basis of agriculture. I believe the Gloucestershire branch has tried, but such appeals have fallen upon deaf or indifferent ears. Now it is different.

Before the system of price support there was a total of more than 500,000 holdings of all sizes. Today there are only 200,000—and overwhelmingly, it is the small farm that has gone. Price support, which has given so much help to the big man, and enabled him to grow bigger, has done little to save the small farm. It is worth noting that as the number of holdings has decreased, the cost of support has risen.

As applied in this country, the EEC system has certainly discriminated against the small farm, to the benefit of the large one—yet on the Continent, the small farmer has reaped rewards, too. Any proposal to redress the balance here appears to have met with an unfavourable response from Whitehall and NFU headquarters, and though the latter has made sympathetic noises lately in favour of the small farm, it seems to me to be charity for the little man so long as the big operator is not deprived of any of the benefits to which he has grown accustomed. What the small farmer really needs is not charity, but a fair deal.

As for recession, many country people have suffered from it for several years past. Smallholders, market gardeners, small farmers and farmworkers have been driven from the land, and rural craftsmen and small industries have been forced out of business. Work which rightly belonged to the country has been taken away.

Country people have had to travel to town to find work, if they can, and in the process they have lost their age-old link with rural life, ways and culture. Many villages have become mere dormitories and lost their old close-knit sense of community. Village schools, shops, post offices and pubs have closed, public transport and other services have dwindled or ceased, and the resident parson has gone in many cases, along with the village policeman. The vital spark of rural life has almost been extinguished.

I am quite certain that the effect has been worse in districts where the small farms have gone—gone to be amalgamated, to make big farms even bigger. Huge, highly-mechanised units have turned friendly field after field into a prairie-like landscape, and husbandry has been replaced by impersonal industrialised farming, more concerned with the production of money than food.

Is it really so efficient, I wonder, when it supports so few people and needs so much support?

Ill fares the land, to hast'ning ills a prey,
Where wealth accumulates and men decay.

Perhaps we should listen more to the poets and less to the economists.

# Waxing lyrical on the wretched root

## April 1985

Good Friday was the traditional day for planting potatoes in the years when most countrymen had little time for leisure, and what time they had was usually spent in the garden. Each gardener had his favourite variety, the virtues of which would be challenged by another, but really it was all a question of taste; some liked a waxy potato, others a floury one. A question of soil, as well; a variety which was excellent in one garden could be very poor in another.

A friend of mine swears by *Sharpes Express*, but in my garden it was useless, and when cooked it went 'all to flop' and tasted horrible. I used to grow the early *Epicure*, which recovers quickly from frost damage, has a good flavour but is rather an ugly shape. *Home Guard*—the name dates the time of its introduction—is a handsome early and a good cropper, but with us it has a disappointing flavour.

For years I have grown *Arran Pilot*, and shall this year, but only a row for the sake of old times. Last year I tried *Sutton's Foremost*, which proved superior, especially in flavour. Again, it depends on the season; the dry summer last year may have suited it, and a wet one could bring different results.

Some years ago I tried *Pink Fir Apple*, an ugly little potato classed as a salad variety and with a reputation as a poor cropper. We use it as an ordinary potato, and with us it crops heavily and is a winner for flavour. In growth it is a rambler, and needs much wider spacing than other varieties. It is not a potato to store, and it must be scraped or scrubbed, since its ugly shape makes it difficult to peel. Seed potatoes are hard to find, but Winfields of Gloucester stocked them this year.

The *King Edward*, when introduced in 1910, sold at half a guinea per seed potato. The heavy cropper *Majestic* was introduced a year later, and both these maincrop varieties are still popular, in spite of the introduction of many new varieties in recent years.

The potato was introduced to this country in the late sixteenth century—reputedly by Sir Walter Raleigh, though there are doubts about that, now. For a hundred years its cultivation was mainly confined to the grounds of the gentry, but by 1700 it was to be found in most gardens. Jethro Tull experimented with it as a horse-hoed crop, yet it was almost another hundred years before it began to be a farm product. With increasing industrialisation, more potatoes were grown on farms to feed the growing urban population, but much of the crops was used to feed cattle and pigs.

According to William Cobbett, who had a poor opinion of the vegetable, there was a 'grave discussion in Parliament about potatoes' in 1800. Someone proposed a law to encourage their growth, and it was, he said, a period of mania for 'this wretched root,' 'this base crop.' His abuse of the potato brought him abuse in return.

Only the onion, perhaps, rivalled the potato in the cottage gardener's affection. To grow large onions was the mark of proficiency. Like most gardeners today I grow onions the easy way, by planting sets, and find *Sturon* the best. Yet it is a tedious task, pushing all those small sets into the ground, only to see many displaced within a few days. Birds are often blamed for this but the real culprits are worms who try to pull the strawy tufts into the ground.

Tomato varieties come and go. I have tried many kinds, including the popular *Moneymaker*, which crops, but is practically tasteless. In my mind *Harbinger* is still best, especially for flavour—it is advisable to grow it from seed, as it is not readily available as a plant.

For lettuce the comparatively new *Buttercrunch* is hard to beat—a cabbage type which 'stands' well. Its tightly-folded, crunchy, wrinkled leaves remain crisp and fresh and have a good flavour; certainly it is worth persevering through the outer leaves, which are tinged with brown and give it a rather off-putting appearance.

# Light as dreams, tough as oak

April 1980

April: the chatter of chain harrows, the ring of the roll, the sweet smell of grass; the cows grazing in the meadows. Spring, we hope, is here.

But already we shall be making preparations for winter; half our year, or so it seems, is spent in growing fodder for the other half. Cattle will be turned out or kept out of some fields, the fields we shall shut up for mowing—'haining,' my father called it. Hain is an old word rarely used in this context today, one of many, now falling out of use, that were the everyday currency of my father and other farmers of his age.

A hay rack or manger was a cratch, a hay loft a tallet, a gap in the hedge a glat or shard. Wallflowers were gillies, buttercups crazies, wood pigeons quists; a quist could also be a peculiar person. Blow, meaning blossom, is still used occasionally, and fall—the fall of the year, fall of leaf—has returned from America. Other supposed Americanisms, too, are old English words which went out of use here but were kept alive across the Atlantic.

Bales of straw have replaced boltings, loosely tied after the sheaves had been threshed. 56 *lb.* of old hay or 60 *lb.* of new made a truss, cut from the rick with a hay knife and tied with bonds of twisted hay.

A strike was a bushel—but before that the strike, I believe, was the stick which levelled off the corn in the bushel measure. The other day the company in my local, the Junction, which included another farmer and two feed merchants, had no idea how much a bushel was; and they all thought it was a weight, rather than a measure by volume.

The bushel varied; the Monmouth bushel of wheat weighed 80 *lb.*; we used to put 2¼ cwt. of wheat in four-bushel sacks. The sacks, incidentally, were always Gopsill Browns, hired from the firm of that name who had a warehouse at Gloucester Docks.

Long rope reins, usually of cotton, were gee-o-lines; they were used when horses worked in long gears, as when ploughing. Badkins, bodkins, whippletrees, whippance, swingletrees—the name varied with the district—were iron or wooden crossbars pivoted in the middle and attached to the hake of the plough. Short gears were used when a horse was working in shafts, and the chains which than took the strain were called tugs. *Captain* and *Duke* were popular names for horses, except when they refused to pull or when they kicked; then they had a variety of them, most of which began with 'B'.

A collar, harness apart, was a large nut used as a washer, to make a long bolt serve the place of a shorter one. This type of improvisation was known as bodging. Bodging was and still is one of the essential arts of farming, though I don't expect it is included in the curriculum at agricultural college. To bodge also meant stopping a shard. (See my third paragraph.)

Bannut was a walnut; cagmag was poor-quality meat; ellum for elm, quick for hawthorn. The grubber was the workhouse, nogman a stupid person, no bottle meant no good. To gule was to sneer; teart was sharp, in the sense of painful; a small load of hay was a jag. The extensions fitted to a waggon when loading hay were thripples or dripples; now they're usually called ladders. Oonts were moles, craturs other creatures, usually people. A lagger was a footpath, a slinnock a strip of land, a yoke a day's work for a pair of oxen. *You English words*, wrote Edward Thomas, *light as dreams, tough as oak, precious as gold.*

Some of our newer words or phrases seem designed to obscure more than enlighten, and are devoid of charm. The computer has bred a host of them, while 'in this day and age' or 'at this point in time' seem to stem from some new brand of pomposity.

In the old days, finish meant finish, but I have my doubts about finalise; when anyone says that a job is not finalised, I suspect it never will be finished; the word 'job' would be replaced by another, longer one, too, yet during the war, Winston Churchill used both 'finish' and 'job' in one sentence to great effect.

Old words are often misused to mislead. *When I use a word*, said Humpty Dumpty, *it means just what I choose it to mean.* But the devaluation of our words and language is a serious matter, far more alarming than the devaluation of our money.

So I have got very tired of escalate, ethnic, pragmatic, traumatic, moderate and their like; they meant something once, but do they any more? As for 'viable;' from its usual current usage, I can only conclude that it refers to a concern that is about to go bust.

# Cleo's glorious day as a rolling stone

## May 1980

Our farm buildings adjoin the road; most of our fields are reached by it, and when we turn the cows out to grass for the first time in the spring someone has to be stationed at every turning.

The cows know what's afoot from the moment the yard gate is unlatched; they sense grass and freedom, and the docile matrons of yesterday are transformed into skittish youngsters. Heeding little, stopping at nothing unless checked in time, they gallop on. 'Who cares? We don't. Anything for a lark.' That's their attitude.

I place my helpers as carefully as any cricket captain deploys his field. This is one day when I'm fully aware that our road leads to everywhere in England and Wales. A false move, a point forgotten and unguarded, and those dervishes could be anywhere.

The squeal of brakes, the clash of cow against car, blood, ambulances, police; in my mind I hear and see it all. Nothing except accident or exhaustion would stop them, and on this day I don't think they'd exhaust very quickly.

Even Cleo, Auntie Cleo, normally so slow and peaceful, will contrive to gallop with the best of them, and even become the leader.

With calves, which go out at the end of May, it's different at first. Having spent their short lives indoors they don't know freedom or grass, and it's a job to get them outside. They do not understand the big outdoors, that tent of blue overhead; and what's more, they don't want to.

But once in the field, after a momentary daze, they do not understand restriction, either. Hedges and fences mean nothing; if they can see through, they think they can go through—and they usually do. For us, the rest of the day is spent in searching, chasing and gathering.

Thomas Tusser died four hundred years ago this month. Although his name may not be on everyone's lips, many of his words are.

Tusser, born in 1524, was a chorister at St Paul's and Norwich Cathedrals, and musician to Lord Paget, but he later turned to farming—'My musick since hath been the plough'—and it is for this that he is still recalled today.

He began to instruct others in the arts of farming, writing in doggerel. In 1557 came his *One Hundred Points of Good Husbandry*; sixteen years later this was expanded to *Five Hundred Points*, with a further *Five Hundred Points of Good Huswifery*.

Much of his advice is still sound today, and many of his phrases are now deeply ingrained in the language: *Look ere thou leap. The stone that is rolling can gather no moss. Naught venture, naught have. Feb, fill the dyke. March dust to be sold, worth ransom of gold. Sweet April showers do spring May flowers. Cold May and windy, barn filleth up finely. At Christmas play and make good cheer, for Christmas comes but once a year.*

But, alas, poor Tusser could not follow the sound advice he gave so freely to others. A rolling stone to the end, on May 3, 1580, he died in prison as a debtor.

# Death of a giant

May 1983

The big old hornbeam tree, with its short bole and massive branches some eighty feet high, stood in front of the house. In winter it sheltered us from the north-eastern winds, and in summer it cast a dappled shadow upon the road. Hornbeams are not common in this district, and I do not suppose many people even knew what ours was; they probably thought it was a beech, which it resembled in many ways.

Hornbeam timber, very hard and much stronger than oak, was used for cog wheels, bobbins, mallets, pulley blocks, screws and piano keys, threshing floors, butchers' blocks and yokes for oxen. When burned it produces more heat than any other wood, and was reputed to make the best quality charcoal.

With each passing year I grew fonder of our hornbeam. In summer, especially after a drought, I would stand beneath it, listening to the slap of rain on its leaves, and watching the drops sinking into the thirsty earth of the lawn or bursting upon the road. In winter I liked to stand beside it and gaze at the intricate and delicate tracery of its twigs against the sky.

I came to regard that tree as a friend, a comrade for all seasons, a talisman. Yes, I was fond of our hornbeam, and all the more so because of that ominous hollow slowly developing where bole met branch. Many times I patted my old friend's smooth bark and fervently hoped that it would at least outlast me.

Fond wishes and fervent hopes are not protection from age and tempest, though. This year, on the first day of spring, a great storm blew up, rattling the rooftops and making the neighbouring poplars sway. The poplars were young and resilient, but my hornbeam was not, and one of its great boughs began to split away from the bole.

At any moment it could have crashed down on the highway. We blocked the road and telephoned the police. Soon an officer arrived and the road was officially blocked. A man from the county highways department came, and an engineer to dismantle the telephone wires, as well as a couple of workers from the electricity board to keep watch on their cables.

A stranger came uninvited to inspect our tree and told us exactly how he would tackle the problem—and then left, as suddenly as he had come, only to be replaced by another, equally talkative and knowledgeable. After telling us how easily and skilfully he could deal with the problem he, too, left.

'We often get that sort,' said the policeman. 'The best thing is to ignore them and hope they go away. They're always the same, have a lot to say—but useless. If they don't go away they're a hindrance, and then I've got to take a firm hand with them. It's a pity about your hornbeam. It's sad when a lovely old tree like this has to be felled.'

A felling expert arrived with two chainsaws, a lithe young man who said little, unlike our two unbidden and self-appointed experts. In a workmanlike way he set to work on the hornbeam, and within ten minutes it was lying in state across the road, its veins full of rising sap and buds about to spring into leaf.

'Death of a giant,' someone murmured.

'All those years of growth, and that chainsaw did for it in ten minutes. It's quite frightening,' said the policeman.

The young man quickly dismembered the fallen tree, and we carried pieces off the road; my family and I lamented the death of a noble and beloved friend; an old woman of our acquaintance drove slowly past and simply said: 'Ha! Ha!'

The feller, the policeman and the others departed, leaving us with a tree stump, a gap in the sky and another in our lives. There was only one thing to do; today, beside that stump, stands a sapling hornbeam, securely staked against the winds.

# A new dawn for small farmers?

May 1985

May used to be the time for root-hoeing; first the singling by hand-hoe, and then going between the rows with a horse-hoe. The hand-hoeing could be back-breaking and tedious, and the horse-hoeing tiring. I can remember my father's injunctions to keep your back bent, to turn the horse quickly at the end of the row, and not to let it tread on the tender plants. Home from school, I'd be told to hurry with my tea; there was horse-hoeing to be done.

Those days and evenings of hoeing by hand and horse, until the leaves of the roots covered the rows and smothered the growth of weeds; of course I was too young then to enjoy the work, and unable to appreciate its quietness. Tractors stood silent in sheds during the whole month of May.

Nowadays few farmers grow roots for their stock—although there is something of a revival in fodder beet-growing—and modern methods have done away with hand-hoeing. The seeds are planted singly and spaced with a precision drill, and the weeds are killed by sprays—in theory at least. The quietness of the country in May is shattered by the noise of machines by the end of the month, now silage has replaced roots as stock feed.

Silage, of course, is not new, but it did not become popular until the advent of labour-saving machinery. Forage harvesters chop the grass like lawn-mowings and blow it into a trailer, and need powerful tractors to drive them, more powerful tractors to put the grass in the clamp, or an even more elaborate system of tower silos.

Silage-making is now general on livestock farms—quick and easy, expensive and noisy. Silage is all right in moderation but the craze for it, like so much else, is partly the result of propaganda and salesmanship. All this fashion for a so-called efficiency which creates more problems than it solves: get farmers onto a new system, and there is all manner to sell, tractors, equipment, fertilisers, sprays, fuels, and then more tractors, bigger tractors, machinery, equipment ... There have been incentives, grants and allowances to encourage farmers to buy labour-saving equipment; these have led to the necessity to produce more and more to pay for them; and this, in turn, leads to the need for more fertilisers.

So we have propaganda, progress, productivity and pressures. We also have overproduction and unemployment. Overproduction strikes me as misplacement and mismanagement in a world where a third of the population is underfed—especially if, as I suspect, the over-production is at the expense of the unfortunate third.

In this country less than 3 per cent of the working population is now engaged in farming; the average in the rest of the EEC countries is 7½ per cent. If 7½ per cent were employed on British farms it would mean another million jobs; think of that, all you politicians. We see the countryside being destroyed as farms grow larger and machines become more numerous, and there is concern about the effects of herbicides, pesticides, fungicides and nitrates in the water caused by the large use of fertilisers. Our rural districts become dormitories as employment decreases in the countryside; rural life becomes devitalised, communities grow moribund, services decline, public transport disappears and shops, post offices and schools inevitably close.

A move to small, more labour-intensive farms is one of the more obvious and desirable ways of reversing the trend. Until recently nothing has been done; nobody has stated the need and importance of the small farm, or its part in the nation's well-being.

# Half century

May 1977

When I was very young, my grandfather said: 'Birthdays are nothing to get excited about, my boy, they come too soon and too often.' At that time, when the spell between one birthday and the next seemed almost an eternity, his remark puzzled me. Now, this month, rather to my surprise, here I am, fifty years old; wondering how and where the years have gone.

But you must admit, there's a ring to it; fifty years, half a century. It just doesn't seem like 18,562 days, though. As for weeks, they seem to be getting shorter; I begin to feel that we really have got a five day week.

Although I've lived all my life in the same parish, when I think back it seems that I lived my early years in a different world. My fifty years begin to seem a long time; goodness knows what older people must think of the changes in their lifetime.

Thoughts of older people bring me to another point: when I was young, age had respect. Not the patronising attitude accorded it today, like calling old people 'senior citizens,' but something quite different. Consequently we felt, as we grew older, that we should be climbing to a peak. But today, with the cult of youth, the peak is reached at around eighteen, which is rather sad; after that, there's just a long downhill run ahead.

There's a great deal of nonsense talked about youth, anyway. It's not all it's cracked up to be. The uncertainties of youth, the miseries of youth; at fifty you don't care if you make a fool of yourself once in a while, and having had a taste of misery you've no desire to wallow in it. Away with Hamlet, bring on Falstaff. And with Falstaff we can say: 'A plague of sighing and grief! It blows a man up like a bladder.'

Life's like riding a bicycle. It's so much easier when you've learned to balance. Sometimes the going is rough and hard, sometimes easy and smooth and then there are those unexpected corners. We come to pleasant spots—fifty appears to be one at the moment—where it would be agreeable to stop and rest awhile. But, alas, this bicycle of life has no means of stopping. We must hurry on to our unknown destination. If we're sensible—if we can manage to be lucky enough—the only thing to do is to make the best of it.

# Rolling clouds and corduroy

May 1982

Adrian Bell, a contributor to this magazine in the 1940s, was the author of more than twenty books that have enriched the literature of rural England.

This month sees the republication of his first book, *Corduroy* (Oxford University Press, Paperback, £2.40). First released in May, 1930, it has been reprinted many times, though it has been unavailable for the past few years.

*Corduroy* is the story of a young man from London who went to learn farming in 1920. In those days, corduroy was worn by agricultural workers, and farming was more or less as it had been for the past two hundred years. The book begins in the autumn, when all was as it should be with farming, and ends a year later with the author taking a few acres of his own.

By then there were signs that all was not as it should be. In 1921 the Corn Production Act was repealed, and soon the price of a sack of wheat fell to ten shillings, a quarter of its price a hundred years previously. The heyday of farming is faithfully recorded in *Corduroy*, and the days of depression come under the spotlight in the succeeding books, *Silver Ley* and *The Cherry Tree*, which I hope OUP will also republish to make this classic trilogy available to a new public.

*Corduroy* begins: 'I was upon the fringe of Suffolk ...' and yet perhaps it was only by chance, and a jar of cider, that that did not read: 'I was upon the fringe of Gloucestershire ...' Yet again, perhaps had it not been for that jar of cider and the clouds rolling over Gloucestershire and the Wye Valley, those wise, elegant and authentic books of rural life would never have been written at all.

In 1920, some weeks before moving to Suffolk, he rode an ex-army motorcycle from London to spend a holiday with an aunt who lived near St Briavels. He retained strong memories of drinking vivid water out of Cotswold rills and rough cider in Gloucestershire, and eating delicious bread and butter at Newnham and an uncouth bacon pie, smelling of woodsmoke, at a Forest of Dean inn. In later years he often told me that the bread and butter at Newnham was the most delicious he had ever tasted, and if I should ever mention Newnham he would exclaim: 'Ah, that delicious bread and butter!'

He arrived at his aunt's house with his motorcycle oozing a mixture of oil and metal. Outside her window he heard the sound of rushing water; he woke in the night and smelled the water like an essence of rocks 'and listened to it gurgling in its course among the ferns.' In Suffolk the water only creeps, and he told me how lucky we were to have brooks that rushed and gurgled. He loved Suffolk, but how he envied us our chuckling streams.

As his holiday drew towards its end he became increasingly reluctant to return to London to start a career in Fleet Street. His father, Robert Bell, was an editor on *The Observer*, and his son Martin is now the BBC's correspondent in Washington.

How, he wondered, had those penurious Lake poets been able to follow their stars? Wordsworth had written his lines about Tintern Abbey not far from the house of his aunt, and when she suggested that he should stop and farm her twelve acres he was tempted.

He engaged a man to help him build a stack of bracken which would provide litter for his stock. The man had bright blue eyes, a straggling beard, and was 'a splendid worker, but ...'

The 'but' came when his stack was half built; the bearded man was found asleep on it, clasping a stone cider jar. He could not be moved, and the stack could not be built over his body, and Adrian Bell simply sat down and watched the clouds roll across the sky.

The next day and the next were the same; the bearded man asleep with the cider jar on the stack, with the rolling clouds above. The stack grew no further, a man came and mended the motorcycle, and Adrian Bell had to return home.

While he had lain by the stack waiting for the bearded man to wake up, he had watched the clouds and thought: 'Fleet Street be hanged.' Within a month he was in Suffolk, and there he remained, until his death in 1980, to farm and to write those wonderful country books.

# The Cobbett of the Twentieth Century

## May 1981

Harold John Massingham (1888–1952) was born in London, the eldest
son of H.W. Massingham, editor of the Liberal weekly *The Nation*. A
townsman who became a countryman, he was a well informed and
prolific writer about all matters rural, a free-thinker who became
a Catholic, a Liberal who became branded as a reactionary, and a
prophet who was dismissed as a harmless crank in his own lifetime.

After Westminster School he went to Oxford and then became
a journalist, literary editor of the *Athenaeum*, and the author of a
number of books which he later dismissed as being tainted with
Bloomsbury intellectualism. He grew tired of daily journalism and
city life, and no doubt his friendship with W.H. Hudson influenced
his decision 'to be off and grow cabbages.'

For a time he lived on the Cotswolds, about which he wrote three
books: *Wold Without End*, *Cotswold Country* and *Shepherd's Country*.
In his later books there are frequent accounts of Gloucestershire, its
people, farming and rural crafts.

He moved to the Chilterns, built a house at Long Crendon,
and started a small rural museum. He wrote a regular column in
*The Field*, contributed to other journals, including *Gloucestershire
Countryside*, and produced numerous books about all aspects of
the country. One in particular, *The English Countryman*, must be
mentioned: it is the harvest of his long experience of and inquiries
into the history and traditions of English rural culture, a scholarly
account of a rural structure now long destroyed.

Most of his books contain warnings about the destruction of the
true source of national life. His words went unheeded, as Cobbett's

did, and it is possible to see Massingham as a twentieth-century Cobbett, though his background and character were different. Both regarded the industrial revolution, the depopulation of the country, the imbalance between town and country and the adoption of false urban values as disaster. Both saw that the price of progress was too high, and neither saw it as real progress, anyway.

He shared with Cobbett the view that real advance was in a backward direction. An old farmer said to him: 'If the new ways won't do, try the old uns again.' Both thought beauty and utility went hand in hand.

Unlike Cobbett, he was never a farmer, although he acquired a sound knowledge of agriculture. Like Cobbett he travelled, noting the state of rural England and talking and listening to farmers and farmworkers and every kind of craftsman. He championed organic husbandry, mixed farming, the family farm and genuine rural crafts—'beauty in use and use in beauty.'

He denounced the domination of machine over man, many modern methods of farming, agri-business, commercial speculation in farmland and 'the enormous, ramifying vested interests which make their profits out of sprays, fertilisers, insecticides and chemical nostra of one kind and another.'

He was all too aware of the political and economic absurdities of importing food while the produce of England rotted, the demise of rural crafts, the takeover of local industries, and the devastation brought on by the Forestry Commission.

He did not want rural England turned into a museum or a playground. He scorned the clicking turnstile, the beauty spots, the arty-crafty, the phoney. He wanted not a quiet countryside but a busy one, alive in the past, not dead in the present. His dream was 'to see England as a whole of which her rock, her wild flowers and birds, her megalithic monuments, her Gothic architecture, her poets, her people and their husbandry were parts inseparable.'

In 1940 an accident resulted in the amputation of one of his legs, but he continued to write and to travel until his death in 1952. The number of his books is impressive, and so is their variety. Perhaps he wrote too much, fulminated a little too often; and perhaps he was before his time, as prophets, by the nature of their calling, must be.

Perhaps he was not always right, either, though I believe that time and events have proved him so on all the major counts.

# Give us this day ...

May 1986

Newly-baked loaves on the kitchen table, golden loaves still gently steaming, the kitchen redolent of them; it is a scene which always gives me a feeling of delight. Every woman high or low, said Cobbett in the days before equal rights, ought to know how to bake bread. These loaves I chance to see from time to time are made of wholemeal flour ground from organically-grown wheat, as Cobbett's would have been.

Then the rich began to eat white bread, and dark loaves became a sign of poverty. Soon, though, with the widespread use of roller mills, the poor began to eat white bread, too, and grew the poorer for the habit.

For more than thirty years my family has eaten wholemeal bread, and often enough has been thought eccentric for it, in the past. Now, of course, wholemeal bread is *de rigueur*, no longer a sign of poverty but of healthful prudence. My wife started baking because wholemeal bread was so difficult to buy, but today almost every baker supplies it.

We use organically-grown wheat because the nasties central to modern orthodox farming are collected and stored in the bran. Some bakers use organically-grown wheat now, too, since people at last seem to be taking an interest in the way their food is produced. All those bland and plausible assurances that everything is perfectly safe are not taken on trust any more—and with good reason, when you consider the times we have been told that things are safe or harmless, and then have been warned that they are not.

There are surpluses of wheat, yet the demand for the organically-grown product cannot be met. It is the same with other foodstuffs.

It seems that some consumers, at least, are ahead of some producers, and if we farmers take heed we could solve many of the problems confronting us. The customer, it used to be said, is always right, and it begins to look as if the truth of that half-forgotten saying is being recognised again.

I stand and stare at those nine new loaves—my wife always bakes nine at a time, some go into the crock and some into the freezer. Traditionally, of course, it should be thirteen, but she has only nine tins, or no room for more in the oven.

I find each loaf a little work of art, and of wonder. Give us this day our daily bread, this stuff of everyday life, so simple yet so profound. Each grain of wheat which went into its making encapsulates the primitive mystery of making things grow out of the earth, dry and apparently lifeless things which contain the means of both their life and ours.

Planted in the cold autumnal ground instead of rotting they put forth a leaf, tender but tough enough to stand the harshest rigours of winter. The snow often comes as a protection, and a nutritious one I believe, but this February past there was no blanket of it and the biting frost and wind blackened and shrivelled the blades of corn. A passer-by would not have supposed that the wheat had been planted, so bare our field looked—yet within two or three days of warmer weather the bare field was transformed into one of green leaf.

A few weeks more, and after a rolling to firm the earth which the frost had lifted, the wheat was harrowed, tearing out some of the springing leaves and covering others with loose soil. Such apparently savage treatment only encourages it to tiller—to put out more shoots while letting air into the soil and creating a tilth to encourage the crop further and deter weeds.

By this month it should be growing apace—or perhaps not, since there is a saying:

> Look at your corn in May
> And you'll come weeping away;
> Look at the same in June
> And you'll come home singing a tune.

The scientist can breed varieties which bear heavier crops; the chemist produce fertilisers to boost yields and sprays to kill weeds and pests; the engineer invent harvesting machinery; but not one of them can touch upon the mystery locked in those tiny grains— grains which look so simple against the complex machines which harvest them.

# Antipathy

May 1978

Some people are frightened of the dark; others are frightened of spiders, or mice, or wasps. I'm frightened of forms. The sight of them causes trepidation; even the simplest ones fill me with dismay.

I'm suspicious of them because I feel they're trying to catch me out, to lure me into a false statement. Then, at some inopportune moment, my licence, my insurance, or whatever, will be discovered to be null and void, or I shall be liable to all manner of penalties, listed in small print. The dreaded 'they' will pounce.

I hate them, too, because there are so many of them, an intrusion on our privacy, our freedom, and our dignity—such as we may have, as human beings. They can be a means of degradation, reducing us to codes or numbers. The post code is an example. We are told that the post code will speed and cheapen postage. Of all the excuses that could have been foisted on us, this is the most patently absurd; we all know the old fashioned way was quicker—and cheaper.

The worst of the questionnaires are those which demand a straight 'yes' or 'no'. You and I, with some experience of real life, know that there's many a time when neither is the correct answer, but the compilers of most questionnaires appear to exist in some rarified world where everything is black or white. So we, their victims, sit with pens hovering. At last we decide on 'yes,' but hesitate; perhaps 'no' is better after all? Suppose we put 'yes' and 'no?' We would like to be truthful and avoid unpleasantness, but like so many whose declared objective is to reach the truth, those forms only blur the issue.

The Ministry of Agriculture is my worst tormentor, with its quarterly census and a few extra ones for good measure. Sometimes its forms have almost three hundred questions. Some of them, like the ones, in my case, about horticulture, can be ignored, though not entirely. There may be questions about fruit trees hidden among them, and I must attend to these, even though I've answered identical queries only a few weeks previously.

The Ministry asks for answers about cropping in hectares. Now I'm sorry, I don't wish to be uncooperative, but I farm in Gloucestershire acres, always have done, and plan to continue to do so in the future. To erase acres and foist hectares on us may be fine in the Ministry's eyes, or it may be just a whim, but whatever it may or may not be, it certainly strikes another blow at the character of the British. Hectares indeed! If the Ministry wants hectares, it can do the converting itself. To hell with hectares!

This quarterly census is important; unpleasantness will follow if it is not completed, and completed truthfully. It is essential, so that the Ministry may know the state of farming and production—and yet it doesn't prevent the Ministry from being taken unawares.

Of course there is the weather—far more important than the Ministry or its forms—and no Whitehall official can interrogate that. As for the official weather forecasters, they seem, to me, a kind of expensive breed of racing tipster. At crucial times, like hay or corn harvest, I make a special point of avoiding their predictions.

Now that I've unburdened myself I feel better, and braver. In fact, I've almost forgotten my nightmare of half the population compiling forms and the other half filling them in, and everybody too busy to do anything useful. It *is* a nightmare, isn't it? Let me know what you think—in triplicate of course.

# The Red Poll

June 1982

After an absence of fifteen years, Red Poll cattle are returning to the Three Counties Show this month. In the past twenty years there has been a dramatic decline in the number of them in this country; with increasing specialisation in agriculture, dual-purpose breeds, and the mixed farms to which they were ideally suited, which have gone out of favour. The result is that Red Polls are now seldom seen, yet they are still not rare enough to be fashionable, as with Gloucester and Longhorn cattle.

When I started to keep Red Polls, almost thirty years ago, they could be seen at every major agricultural show, and at many minor ones. At both the Dairy Show and at Smithfield Show they were present in force, putting up a creditable performance against the pure dairy and the pure beef breeds. But a breed which combines milk production with beef production cannot reasonably be expected to triumph over single-purpose breeds in their respective fields, and although the Red Poll has produced its quota of two-thousand-gallon cows, its milk yields tend to be respectable rather than spectacular.

Red in colour and naturally hornless—hence their name—the cows are hardy and thrifty and noted for longevity. Some of ours are still breeding and giving a thousand gallons or over at twelve years of age or more; our oldest will be eighteen this month, and is due to calve a few days after her birthday.

There used to be several Red Poll herds in Gloucestershire, some of which were world famous. The Seven Springs Herd regularly produced Smithfield Show champions and exported cattle overseas. Also notable was the Lydney Herd, founded in 1921 by the first

Viscount Bledisloe, whose keen and practical interest in farming earned him the name the Grand Old Man of Agriculture.

At one time there were so many titled owners of Red Poll herds that the breed was known as gentleman's cattle; they looked well in the park, and, hornless, they were no danger to bloodstock. Such a reputation did not do the breed any good as far as the general farmer was concerned, and on paper, at least, a thousand gallons did not look as good as the fourteen hundred of other breeds. This is why, by and by, the breed dropped out of favour.

Most of the native breeds of English cattle were once dual purpose, until the improvers of the late eighteenth and early nineteenth centuries turned them into beef breeds. There are no native dairy breeds of England; of those we now have, two came from the Channel Islands, one from Scotland, and two from Holland—one more or less directly and the other via Canada. Now, even our native beef breeds, upon which we prided ourselves and exported all over the world, are faced with strong competition in this country by the Continental breeds. Once the slogan was 'Britain can breed it;' now it seems that Britain can import it.

Norfolk and Suffolk breeds were possible exceptions. The dark red and horned Norfolk was mainly beef; William Marshall's *Rural Economy of Norfolk*, 1787, spoke highly of its meat qualities and its hardiness. The hornless Suffolk, brindled, dun or red in colour, was renowned for large yields of rich milk; Arthur Young, the first secretary of the Board of Agriculture, was much impressed by it.

Robert Bakewell, the famous cattle and sheep breeder, thought the Suffolk an ugly little animal which would look as well upside down. She may have survived if the railways had come in time to transport her milk. Instead, the Suffolk became one of the ancestors of the Red Poll.

At the beginning of the nineteenth century, John Reeve, a tenant on Coke's Holkham Estate, crossed the Norfolk with the Suffolk with the intention of combining the hardiness and beef of the one with the milk of the other. By about 1840, a new dual-purpose breed had evolved, and gradually it spread outside its native East Anglia to the rest of Britain, to Australia, New Zealand, South Africa, Kenya and the Americas.

It, too, became the ancestor of new breeds: in Brazil, Jamaica and Columbia, the Red Poll has been used on native cattle to produce the Pitangueiras, the Jamaican Red and the Velasquez breeds.

It is ironical that a breed which has shown that it can thrive in most climates of the world has not flourished in the climate of present English agriculture; but fashion, like our weather, is notoriously fickle.

# Newnham on Severn

## June 1982

It is often called a village now, and there is even a shop which calls itself The Village Stores, but I can remember a time when its inhabitants would have been most indignant at such talk. Newnham, they would declare, was a town, with a town clock and a town hall, though they usually called the latter 'The Comrades.' The town clock and tower still stands, and has recently been renovated, but the town hall is now the Club. Perhaps, strictly speaking, Newnham *is* a village these days, but I for one shall go on calling it a town.

Newnham was once one of the five boroughs of Gloucestershire, the county's only one west of Severn, and its status as such went back to 1187. Kings came here: William II, Henry I, Henry II, Edward II, Edward III, some of them several times. John used to visit Flaxley, but there is no record of him at Newnham. He is reputed to have presented it with a Sword of State, but though the sword is real enough, and is said to be the second largest in the kingdom, it is thought to be of a later date than John.

Henry II and his Chancellor, Thomas à Becket, signed a document in Newnham, and later Henry II met Strongbow, Earl of Pembroke, here, and gave him a wigging for allowing himself to be made King of Leinster. They then patched up their quarrel, and together set sail to conquer the whole of Ireland.

A regular market was established in the town by the twelfth century, but six centuries later the poor condition of the roads had stopped the trade in corn, and the rest of the weekly market became little-used. There also used to be two regular fairs: on St Barnabas Day, June 11 and St Luke's Day, October 18. By the nineteenth century they were mainly fairs for the sale of horses and cattle, becoming partly pleasure fairs in the early years of this century.

During the First World War the sale of livestock ceased, and the fairs became pleasure fairs only until the late 1920s, when they disappeared altogether. Newnham Fair Day, the one held on St Barnabas Day, was the traditional one for the local farmers to start mowing grass for hay. I have been told: 'The old uns ud bust their guts, sun or rain, to start on Newnham Fair Day.'

There was a market house. This and the market stalls were probably situated in the middle of the High Street, where there are now lime trees and daffodils. The width of the High Street has led to the supposition, often presented as a fact, that there was once a row or even a double row of houses on this site.

If this were fact, it is surprising that no foundations have been found, and until they are it can only be regarded as a plausible but doubtful theory. Another common belief, that there was no outlet at the northern end of the High Street until the middle of the nineteenth century, is disproved by contemporary maps.

The lime trees were planted during the nineteenth century, and the daffodils in the 1930s. The latter are Newnham's pride, and have earned it the name of Daffodil Town, though because the grass cannot be cut until the leaves have died down, the brief period of glory is followed by a longer one when the High Street looks unkempt—the morning after the night before.

Shipbuilding was one of the town's industries. In 1776 a ship of four hundred tons was launched; called the *Nancy*, it was built for a Bristol merchant, and was the largest to be built on the Severn until that time, though one of six hundred tons followed two years later. Newnham's port traded in timber, coal, oak, bark, cider, hides and glass.

Tanning was important, too, and Newnham was a centre of the industry, with the necessary oak bark easily obtainable from the Forest of Dean. Although the town has little of the Forest atmosphere or influence about it, much of its former importance and prosperity were linked to its hinterland. Its port and most of its industry depended upon the products of the Forest; the glassmakers came here because of the coal, and the town's glassworks were in all probability the first to use coal instead of charcoal. No trace of them remains.

Tallow-making was another of its industries, one instance at least of products going into, rather than coming out of, the Forest, for its candles would have been used in the Forest's iron mines and coal pits. Forest iron would have been used to make nails at Newnham—and Nailyard is still a place-name here, along Station Road.

The railway station opened in 1852 and was closed in 1964; now there is a campaign to have it reopened—the strong feeling being that it should never have gone in the first place.

Weaving, dyeing, felt-making and rope-making were other former trades; printing still survives, as well as such usual country skills as saddlery. As with the markets, business declined, and the port lost trade with the opening of Bullo Dock in 1809. As old photographs show, there was still activity on the river, and well into this century the town continued to be the centre for a considerable district.

The Bear, at the junction of Passage and Back Lanes, was a posting house, and recorded in 1637. In the eighteenth century the borough and manor court was held there, and in the next century petty sessions. John Byng, author of *The Torrington Diaries*, visited it in the late eighteenth century: 'I entered the Bear Inn at Newnham, with a good appetite, and found a round of beef just taken from the pot, which I strove to devour, and likewise a gooseberry pie. It is always my rule to stop (if possible) about noon, at second-rate inns, and take the family fare; as one commonly dines much better in that way, and at half the expense of an order'd dinner.'

On the route to South Wales, with a river passage, port, markets, fairs, industries and commerce, it is not surprising that Newnham had a number of inns and alehouses. Some are now private houses, while the Anchor, and perhaps others, have vanished. They were not all in business at the same time, but there were about a dozen at the middle of the last century; eight at the beginning of this century, the Victoria, True Heart, Ship, Lower George, Railway, New Inn, Upper George, and Britannia; and today there are only three.

The town hall was built in 1849 and housed the county court. The town clock was erected by public subscription in 1873. The castle is something of a mystery, for no trace of it remains. As far as I am aware, the same applies to the prisons of the thirteenth and sixteenth centuries, and the theatre of the eighteenth century, because for all its history, Newnham has few reminders of the past.

Some have been swallowed by the river or destroyed by fire—including a mysterious Great Fire of around 1300. One of Newnham's churches was a victim of water and another of fire, and the present one is at least its third.

Rebuilding in the eighteenth century and Georgian facades have helped to disguise the age of the town generally. Today the High Street has much the same appearance as it had in its earliest photographs, but in this century there have been important invisible losses of trade and autonomy, not least that of the Newnham Urban District Council in 1935. Once the town had a fierce pride; perhaps it still has, except it does not shine so brightly.

It certainly retains its solid air of respectability, and it is hard to believe it was ever anything but respectable. Even Newnham has had its turbulent moments, however, in addition to the perils of water and fire. At one period the town was subject to raids by bands of armed men from the Forest. There was the unpleasantness during the Civil War, when one of Winter's men fired a barrel of gunpowder in the church. The explosion blew Royalists and Roundheads out of the building without killing any of them, but Massey's men, understandably enraged by such an unsporting act, set upon the Royalists and slayed many of them.

In the nineteenth century Newnham, more than anywhere else in the district, was vigorously opposed to Nonconformity. Windows were broken at the house where a Nonconformist meeting was held, and the family found it prudent to quit town. Later a chapel and its contents were wrecked on several occasions, and the town was taken to court and ordered to pay for the damage.

Some time during the nineteenth century 'certain persons of consequence in Newnham' became 'rather elevated' after a wedding, and either threw into the river or made a bonfire of seats from The Green, which were fashioned in the shape of 'queer beasts.' A reward of five pounds was offered for information, the resolution being voted by the culprits themselves, but the reward was never claimed.

A well-known businessman of the town, whose numerous activities included the purchase of bad debts, was once set upon by some people who he had dunned; and every native of Newnham over fifty will be familiar with the oft-chanted phrase: 'If you vant to buy a vatch, buy a vatch, and if you don't vant to buy a vatch, take your dirty face away from my vindow!'

Older people may remember the embarrassment of the couple who had to ring the church bells in order to be released from the belfry; or the dentist and his wife who travelled from Gloucester in a bath chair, each taking a turn to ride; all, no doubt, very enigmatic to the uninitiated, but proof that the town was never quite as staid as it appeared.

All things considered, it is surprising that Newnham has not figured more in the diaries of travellers of yesteryear, or in modern topographical books. Rudder in the eighteenth century called it a flourishing little town. The Reverend F.E. Witts in the nineteenth century saw it as a dull, small town on a hill above the Severn. Sedate perhaps, but dull? Was Witts in a bad humour on that July day when he passed through Newnham?

In the middle of this century, more accurately and happily, F.W. Baty called Newnham the Lady Who Never Grows Old. Now, thirty years on, she remains just as ageless, and one of the most delightful places, town or village, east or west of Severn.

# From suns to June moons ...

June, when the cuckoo changes its tune, and dandelions are transformed from suns to moons. June 11, St Barnabas Day, the day of Newnham Fair, was the traditional start of the season for mowing grass for hay. Now they start earlier, and they mow for silage, not for hay. During the past decade or so silage has been fashionable—a fashion, incidentally, that coincides with the advent of expensive silage machinery. But then, the trend in farming is often that way.

Silage-making was introduced into this country a hundred years ago. A pit silo, which still exists, was built in Herefordshire in 1885, and is probably the oldest in the country. Fifty years ago tower silos, made of brick, concrete, steel or timber, sprung up all over; but for one reason or another—mainly the labour involved—silage-making did not become popular.

## Less labour

After the last war the invention of the buckrake—a near-copy of the haysweep—lessened the labour element, and helped the practice become more general. The hay-sweep was invented in the middle of the nineteenth century, as was the horse-drawn mower, which gradually replaced the scythe. The horse-drawn swath-turner appeared about a hundred years ago, as, more surprisingly, did the hay-baler. The pick-up baler we know today came from America, and was in general use here by the early 1950s.

Grass and clover are, of course, native, but many species grown today were introduced to the country. Red clover, perennial ryegrass, sanfoin and lucerne all came during the seventeenth century, and Dutch white clover in the eighteenth.

If silage has replaced hay to a large extent, it has replaced kale and root crops even more. Turnips came in the seventeenth century, and were later to become an important element of the Norfolk four-course rotation—turnip; oats or barley; clover and ryegrass; and wheat. Hoed during the summer, the turnip was a 'cleaning crop'; it provided winter keep and, when sheep were folded on it, fertility for the following crop, which was undersown with grass and clover. The clover, when ploughed in, provided humus and fertility for the following crop of wheat.

The Swedish turnip, commonly called the swede, came a century later, as did the mangold, which originated in France. The mangold, as Professor Boutflour of Cirencester Royal Agricultural College never tired of saying, was largely composed of water; but an old farmer replied: 'If 'tis mostly water, 'tis damn good water'.

Cow cabbages were once a popular crop, but marrow-stem kale, introduced at the end of the last century, superseded it. Rape was grown two hundred years ago for its oil; later as a grazing crop; and now, again, for its oil. Today, when in flower at around this time of year, it makes much of our countryside look like a Van Gogh landscape with its vivid yellow hue.

Fodder beet arrived from Denmark less than fifty years ago, while its cousin, the sugar beet, was first grown in England in 1909. William Cobbett, who had seen maize growing in America, tried hard to popularise it as a grain crop here, calling it Cobbett's corn. Maize failed to ripen in our climate, and it did not become popular until it was accepted as a silage crop.

# Jubilee

June 1977

'Penn'orth of acid drops, ha'p'orth of aniseed balls, ha'p'orth …' said Tom.

'A sherbert fountain, liquorice …' cut in Ronald.

'No, I'll have bulls eyes, humbugs, an' …'

'An' I'll have toffee an' pear drops …'

The boys spoke at the same time, changing their minds. Farmer Smith had given them sixpence each for cleaning the pig cots that morning: 'Mind, it'll be back to thruppence come Saturday, same as usual; it's only sixpence today 'cos it's a special day.'

'Ha'p'orth of butterscotch,' said Ronald.

'Two penn'orth of barley sugar,' said Tom.

'Make up your minds,' said Mrs Dobbins impatiently, her hands and forearms red and still moist. 'I haven't got all day, I'm in the middle of my washin', and I want to go to the do, same as everybody else. He's been a good king.'

Tom and Ronald were friends. Both went to the little village school, and lived in adjoining cottages. To bypassers these were picturesque, but rather dark and damp inside. Water was drawn from a shared well, lavatories were at the bottom of the garden; electricity was something they had in towns, and the nearest bus stop was two miles away.

'It's all right, ain't it?' said Tom, kicking stones outside Mrs Dobbins tiny shop. 'A day off school, and all that up at the Manor s'afternoon.'

Tom and Ronald were among the first arrivals for 'all that.' 'Coo!' said Tom, surveying the flags and bunting, the marquee and tables on the lawn. 'It's all right, ain't it?'

By half past three all the children of the parish were at the trestle table under the beeches, but even here the air was warm, remarkably so for early May. Some were raggedly dressed; it was difficult to keep half-a-dozen well turned out on the thirty shilling weekly wage of a farmworker, or on dole money.

At the head of the table sat the squire, his old eyes twinkling, though his heart was split between joy and regret. Joy at the sight of the happy young faces round him; regret that he and his wife, Hester, were childless.

Hester sat on his right, the new vicar on his left. He wasn't too sure about this new chap; too prim and earnest and censorious. Not like old Prendergast, now dead and gone, God bless him. Prendergast had liked a day's shootin' and a drop of port, or a mug of cider with the men.

Everyone's friend, old Prendergast; doubt if they'd ever feel that way about this new chap. Still, maybe he'd mellow, mustn't condemn the fellow.

The new vicar rose to say grace, and added a long-winded homily on royalty, patriotism and morals.

'Silly young fool,' thought the squire. 'They don't want that, they want the grub.'

There were pork pies, sausage rolls, ham sandwiches, jellies and blancmanges, fancy cakes and buns, a bottomless well of lemonade, but best of all, there was ice cream; a treat rarely tasted by the children of this remote rural parish in 1935.

'You know what'll happen if you drink too much lemonade, Teddy,' screeched Mrs Gibbs, the schoolmistress.

The old squire was a little overcome by it all, the historic occasion of the King's Jubilee, and all these children. He should have had grandchildren about this age now. And what of the estate? Been in the family for hundreds of years; not much of it left now, with death duties and depression.

He was wondering, too, how the devil he was going to find the money for pay for these celebrations. 'Up to my neck already,' he mused. 'Have to sell another farm soon, what with one thing and another, but who wants to buy a farm these days? Farming's gone to pot. Terrible thing. There were three cars in the village: his own, the doctor's, and the wealthy farming chap's down the road. Sometimes, in his darker hours, he wondered whether that figure might be reduced to two again.

When at last even the hungriest and greediest child could eat no more, the squire stood up. The sun, slanting through the trees, had caught and reddened his face, and his large white moustache twitched. He looked rather like one of those big, elderly, benevolent animals seen in children's picture books. He mumbled a few words, sat down, and then rose again to say: 'God save the King and Queen.'

'Come along, children,' shouted Mrs Gibbs, clapping her hands, the sun glinting on her spectacles. 'Stand to attention and sing *God Save the King.*'

But they were not far into the anthem before little Sally Jones was sick; violently sick all over her new party frock; over the table, and over her neighbours. In the ensuing melee, Willy Cook's jubilee mug was broken. He let out a howl, and seemed like continuing with it for a long time before the squire produced another one, having bought a few extras, just in case. Peace was restored; Willy smiled through his tears like sunshine after a shower.

The games went merrily enough, until Gwenny Sherman fell in the lily pool. She was rescued and dragged dripping and protesting across the lawn.

'I didn't fall, I was pushed,' wailed Gwenny, through the duckweed. 'Ronald pushed me.'

'Ronald, I've got my eye on you,' said Mrs Gibbs. 'I haven't forgotten what you wrote on the wall.'

The Punch and Judy show pleased everyone, but by the time the curtain came down the squire was quite happy to toddle off to the library for a quiet snooze before the evening's gaieties.

At half past six the band arrived and was guided by old William the gardener into a marquee housing several barrels of beer. Twenty minutes later they emerged, wiping their mouths with the backs of their hands. Soon they were playing a variety of patriotic tunes. Most of the parish was now assembled on the big lawn; the women, even the farmers, stiff and hot in their heavy, formal clothes; and the farmworkers, incongruous and awkward in navy serge suits and big flat caps.

Friends gathered in little groups, talking, or listening to the band. The old men huddled beneath a lime tree, talking of animals and crops and deeds of long ago. The great freeze-ups and snows, the dreadful flood and long droughts. The long weary hours, too, of hay and corn harvest: 'The young uns 'ould never stick it today, nor

drink the cider we drunk.' The young men on the fringe listened
with open mouths. One day they would recount this day, and a
few great deeds of their own; mighty crops, splendid animals, and
dreadful weather.

The band was playing old favourites. Everyone agreed there was
nothing like a brass band. Ethel, the kitchen maid at the Manor,
now dressed in her finery, was watching out for Fred, her young
man, who had recently enlisted. At last Private Fred Sherman
appeared, resplendent in his uniform. Arm-in-arm, he and Ethel
stood listening to the band, before creeping quietly into the jungle
of rhododendrons.

The sweet tones of *I'll be your Sweetheart* floated over the park,
the pigeons crooned in the oaks, the rooks cawed in the elms. Mist
rose from the lake and mingled with the scent of leaf and flower and
grass; there was even a touch of magic in the air.

The band played *Rule Britannia* again. George was the sailor
king—*Britannia rule the waves*—and much of the land, a quarter of
the map on the schoolroom wall, was coloured pink. Still, the sun
never set on the Empire. London was the hub of the world. British
goods were the best, and Europe was far away, and there was beer and
beef, ham and pie in the marquees.

Sparks sprayed from the bonfire, rockets pierced the air: 'Look at
that un, he's a bobby-dazzler.'

'Good ain't it?' said Tom.

'Ray, ray, ray!' cried a small boy, kicking himself with ecstasy. The
band played on, people were singing, others were dancing, children
were running round and round. The shadows lengthened, the night
was still warm, but a faint breeze rustled the leaves. The pigeons had
flown to rest, a cow bellowed in the distance, a pheasant called, and
an old white owl flitted silently overhead. The revels were almost at
an end.

The band played *After the Ball was Over* at a slow tempo. Hand
in hand by the sundial stood the old squire and his lady. Ethel gazed
shyly at her Fred; even the vicar was smiling. Old William, who was
worse for beer, belched. Bronze-faced men stood round the bonfire.
Mrs Gibbs hushed the little freckled boy. Tom murmured to Ronald:
'It's been all right, ain't it?'

The bandmaster raised his baton. Everyone stiffened at the
opening bars of *God Save the King*.

# Pub conversation

June 1981

In the pub they were discussing the relative merits of potatoes. Some preferred the older varieties, staunch champions of Arran Pilot, Ulster Chieftain and Home Guard; others plumped for the newcomers, and Joe said you could not beat Sharpe's Express.

'Huh,' snorted the little man. 'They'd cook all to flop.'

'It's knowin' how to cook 'em,' retorted Joe.

'It all depends on your ground, you must grow the taties that suit your ground,' someone else chipped in.

'I've read one of your books,' said the thin man, leaning towards me. 'It wasn't bad.'

'They tried to stop us growing King Edwards,' said the little man, 'but they didn't succeed.'

'But your book wasn't good either,' the thin man told me. 'I could write a much better one. Y'see, I have led an interesting life.'

'Well, why don't you?' I asked him.

The thin man looked at me sharply and said: 'Why don't I what?'

'Write a book.'

'Oh,' he smiled. 'I haven't time. And besides, I can't write.'

'But, you've just told me you could write a good book.'

'Yes, so I could, there's no doubt about that, but as I've just said, the only snag is that I can't write.'

There should have been some neat reply to that, but I couldn't think of one at the moment; and, to tell the truth, I have not been able to think of one since.

'It's all them artificials what spoil potatoes,' said the little man.

'It's been a bad spring for plantin' corn, an' that cold all through April an' snow an' all,' said a farmer.

'What did I tell you?' exclaimed the little man. 'When you was all talkin' about that lovely weather in January I told you we'd all suffer for it later. I told you spring in January 'ud mean winter in April.'

'The price of beer is summat cruel, not many years ago it were two bob a pint and now it's eleven,' the portly man grumbled.

'They make beer out of barley an' the price of barley's fallen an' the brewers put up the price of beer,' muttered the farmer.

'And the tax,' said the portly man 'they keep puttin' up the taxes. Where'll they find the money when they've taxed us all to death I'd like to know.'

'The Common Market tax us on our milk, they take a whole week's milk a year, not just the profit but a whole week's milk, work an' expense an' all,' said the farmer.

'There used to be highway robbers an' now they be gone we've got this other sort,' said the portly man.

'I'd rather have the highwaymen,' said the little man. 'They wasn't robbin' you all the time an' they didn't keep pesterin' you with papers neither.'

'My father said there were two things which spoilt his life, income tax and haymaking,' said the farmer.

The thin man at my side had fallen into a reverie.

'It's all gettin' too much for me,' said the little man sadly. 'What with the taxes and the papers, an' the price of beer an' the funny weather we keep gettin', and the metres and suchlike. Oh dear, I don't know what'll become of all on us.'

'Cheer up,' the portly man advised.

'Cheer up!' snorted the little man. ''Tis all very fine to say that, the cost of cheerin' up is goin' up by leaps an' bounds with beer the price it is.'

The thin man stirred and said to me confidentially: 'I've been thinking.'

'About your book?'

'I've decided the best thing to do. I'll tell you all the interesting things about my life and you write it all down. Mind you, my name will be on it, not yours, because it'll be my book, and jolly good it'll be, too. You'll see that when you've read it.'

# Ways with hay

Hay used to be made into cocks in the field, and pitched on to wagons, and then built into ricks. Then came hayloaders hitched behind wagons, and later the sweeps, pulled by horses, or fixed to the front of tractors, which pushed the hay to the ricks.

A fearsome kind of grab, called a devil, and powered by a horse, or an elevator driven by a petrol engine, saved the arduous task of pitching hay to the rick. The sweep method was speedy and saved the double handling, the loading and unloading, but it meant that the rick had to be built in the field, which was not too convenient in the winter, when hay was needed for cattle in yards and sheds.

Some farms then began sweeping hay to stationary balers, and taking the bales into barns. They were wire-tied, big and heavy. Next came the pick-up balers, which we still have today. In those early days, though, the art of the pick-up baler had not been mastered, and some of the hay was mediocre stuff.

The art of building ricks of bales, too, was no easily-won skill. The products of the stationary baler, though difficult to handle, were solid and easy to build. The pick-up bales, however, slipped and tumbled, fell apart, and ricks collapsed; it was a game, a mess, until we had learned the way.

But cocks, loaders, sweeps or bales, haymaking is hell. It is not, and never was, the idyllic pastime that some would have you believe; people whose livelihood has never depended upon it may see it that way, but the reality is different. There has been too much of that nonsense about farming, from writers, painters and broadcasters, all, more often than not, observers rather than men of the fields.

The pick-up baler and its accessories, the sledge, loader, and elevator, have speeded and eased the task, or at least have enabled fewer men to perform the jobs that formerly occupied many.

Worse than the toil, the sweat and the dust, is the worry. There is always the weather; a storm of rain, especially if it is prolonged, can ruin hay just fit to bale. However much you turn it afterwards, it is never the same, and the worse it is, the more expensive it is to make. We farmers are renowned grumblers, but all in all, we are probably more philosophical than most; we have to be.

# Sinking holes, sinking hearts

June 1984

Fencing is a never-ending job on a farm. No sooner have you made one fence stockproof than another needs attention. Stakes rot, wire breaks, stock escapes—and usually at the most inconvenient times, during haymaking, at night, on wet Sunday evenings. Crops are trampled and ruined—and the animals themselves may be endangered by the very crops they are destroying. There are neighbours to be considered, too; bad fences, it is truly said, make for bad animals—and bad animals make for bad neighbours.

Hedges, in theory, are fences, but in practice only the well-maintained all-hawthorn hedge is really stockproof. Ditches also need fencing, and while electric fences are easy to erect and popular, they are not as satisfactory as a strand or two of barbed wire.

Every spring, then, we try to see to our wire fences—but there is never enough time to do enough of them properly. All too soon it is silage-making time, and then haymaking, and eventually we have to resort to 'bodging'—patching-up and making do with a temporary job, something we hope will serve for the time being. We know this bodging will not last, that it is a waste of time and sooner or later we shall have to do it all again, and do it properly.

A proper job means new wire, a stake every three or four paces, and a straining-post every fifty yards and at every sharp turn. Stakes take time and effort to drive into the ground, especially when the earth is as dry as it was this spring. Digging post-holes and then ramming the earth back in around the timber takes even more time.

Digging the first hole is a lot of effort, slow, irksome work. You think of all the other holes to be dug, and your heart sinks more easily than your spade. The second hole is easier, less irksome, and by the time I have dug half-a-dozen I begin to enjoy the job. Soon I have no desire to do anything but dig holes. Digging holes is absorbing, and what is more, it is quiet. You can hear the singing of the birds, and work at your own pace, instead of the pace of the machinery.

Occasionally my grandson, Harley, comes to help me. First he brought his own spade, a toy one not up to the work, and not good enough for a chap who knows how to handle a tool, in spite of his tender years. On my next visit to town I bought him the smallest and lightest border spade I could find, and with this he can really get to work.

For all his enthusiastic help, though, he is also the cause of hindrance. Every time he sees a worm we both have to stop digging until he has picked it up and put it safely in the hedge-bottom. This reverence for worms is infectious, and even when he is not there I find myself rescuing them.

Then there are pebbles. Harley has to save all the most pretty ones to take home to his small sister, who is very fond of them. She is fond of worms too. She speaks to every one she sees, and on one occasion I saw her make a home for one out of moss. She put it in, and then she gave it the only sweet she had. When the worm made no attempt to eat it she took it back and would have had it herself had grandmother not stepped in.

There I was, saying I had not enough time, and then I found myself rescuing worms and pebbles and discussing life with Harley. Perhaps had I not done so, I might have got a few more yards of fencing done before we started silage-making. But was it a waste of time? Would a few more yards have been adequate recompense for all I gained? It all depends how you look at life; and what you call waste, and what gain.

# Haymaking's game of chance

## June 1986

February, March, April, May, into June; I have never known it so cold, so wet and rough for so long. It has been colder, wetter and windier during these months, but not for such a long, unbroken spell as it was this year. We managed to plant some barley, but most of the fields never got harrowed or rolled, and it seemed as if the trees and hedges would never turn green. Eventually nature could wait no longer, and in spite of the weather leaf and flower came forth, and the birds began to sing and build nests regardless.

The plum blossom was three weeks later than usual, the apple trees did not bloom until the third week in May. The sun hardly dared to show its face, and still the rain fell and the cold winds blew. It must have been a bad spring; I have never heard so many non-farmers expressing sympathy for the likes of me.

As I write, matters look more cheerful; the sun is shining and we have had no rain for three days. Now there will be so much to do, and so much that will not be done. Already some farmers have started silage-making; those of us who have not started shake our heads and say there is goodness in the grass yet, but secretly we wish we had made moves. There is nothing quite so irritating to a farmer, not even the weather or politicians, as a neighbour starting a job before him. We loudly declare that the land is too wet, the grass is not fit, the corn is not ripe, the weather is not right, but how we wish that we had been first away.

Yet these much-scorned pioneers do spur the rest of us on—and with haymaking I find I now need some kind of spur, since each year I view the prospect with added trepidation: the weather to contend with, machinery which has a habit of breaking at the most difficult of times, the long hours, the toil and sweat, the heartbreak and backbreak. That is the game of chance called haymaking.

Methods have changed, but it remains a chancy business. If you cut a lot of grass and the weather is fine you congratulate yourself on your boldness. If it rains you curse your foolishness. The more it costs to make, the worse it is.

The pick-up baler came as a boon and a blessing. At first a lot of indifferent hay was made with pick-up balers, and as bad or even worse were the indifferent stacks of bales. Bales would tumble out, especially at the corners, and parts of the stacks would collapse overnight. We tried tying them together but to little avail; if bales wanted to tumble, as they mostly did, then tumble they would. Until the technique of building stacks of bales was mastered, there were ugly sights and tempers on several farms.

At first we were well content to lift up the bales by hand as they were dropped by the baler. It meant a lot of walking until the sledge was invented. The bales still had to be lifted on to the trailers by hand until the tractor-mounted bale-lifter was invented, and there have been other inventions since then, some more successful than others. In spite of all this though, the toil of haymaking remains, mainly because most of the inventions have lessened the number of labourers in the fields.

Old photographs of haymaking with many men—and often women, too—make the task seem idyllic: the men wearing straw hats, the women with large sun bonnets, the rakes and pitch forks, the stately wagons and horses. What a peaceful scene it seems to all used to the roar and capricious ways of modern machinery.

But what the photographs do not convey is the sweat and toil and long hours, and the time when the sun is not shining. There is no haymaking without the sun even now, with our vast array of machinery, and that is probably why so many farmers have turned to silage. The fact remains that you cannot beat good hay, though—if you are lucky enough to make it.

# The kingdom's richest commoner

June 1986

James Wood, the Gloucester banker, died a hundred and fifty years ago this year, leaving a fortune of more than a million pounds. Reputed to be the richest commoner in the kingdom, he left mystery, as well as money, and as late as 1950 Americans bearing his name were still battling for their share of his fortune. The simple fact was that while he was famed in life for his wealth and eccentricity, Jemmy Wood won even greater notoriety in death through bitter legal wranglings over his will.

Jemmy Wood was born in Gloucester on October 7, 1756, descended from the Wood family of Brookthorpe Court, which had a three-hundred-year connection with the county. In 1716 his grandfather, also James, had established a bank in Westgate Street, one of the oldest private banks in the country, and Jemmy was the last of a family noted for thrift.

He had relatives—I do not suppose any wealthy man is ever entirely lacking in relatives—but he would have nothing to do with any of them except his cousin, Anthony Ellis. He certainly showed his affection for money at an early age. When his father told him to get a penny for a little girl who had brought a bunch of flowers, Jemmy did as he was ordered and said: 'Father, smell that penny. It will smell as sweet in a week hence as it does now. The posy won't.' His father was delighted at this remark.

It is not known when Jemmy succeeded to the Gloucester Old Bank, but by the beginning of the nineteenth century he was a familiar figure standing at the door of the old shop in Westgate Street, his rigid countenance occasionally relaxing into a hard smile at someone passing by. He had a large, prominent nose, a forehead which receded sharply and a mouth and eyes which indicated a love of pleasure, for though a miser, he was fond of good food and drink.

He always dressed shabbily, in old breeches and hose, low shoes with silver buckles and shirt showing beneath his waistcoat. On at least one occasion his faithful assistant, Jacob Osborne, remonstrated with him about the condition of his clothes, and told him to buy a new suit before he went to London. He argued that nobody in London would know him, anyway, and when some weeks later Osborne told him he should be ashamed to be seen in *Gloucester* in such clothes he again had an answer: 'Tain't as if I was going from home, Jacob.'

His base in Westgate Street was an old overhanging half-timbered house with small panes of glass in the windows and a sign proclaiming Gloucester Old Bank. To reach the bank you had to pass through a shop, for Jemmy was also a draper, haberdasher, general merchant and, apparently, an undertaker. At funerals he could weep copiously, but this did not prevent him from selling gloves, scarves and hatbands to the mourners.

To help raise money to fight the war against Napoleon, the Government resorted to a state lottery, and Jemmy was appointed as an agent to sell tickets. Three sold by J. Wood Esq. in 1805 brought each of their buyers £20,000, and Jemmy received a large commission on each of them. Shrewd and careful, he never invested in any risky ventures, so never lost any money. He was always at work, paying strict attention to his bank, shop, property and several farms, and he was unrelentingly hard-fisted, charging large interest rates and giving small ones.

The terms for money on deposit at his bank were per annum, and if cash was drawn before a full year had passed not a penny of interest was paid. On one occasion £800—a huge sum in those days—had been deposited, and when the twelve months were up the depositor asked for its return.

'Certainly, certainly,' said Wood. 'It be all safe; I have got it all right.'

When the money was put on the counter, Wood asked: 'Is it all right?'

'The money is all right, Mr Wood; but you have forgotten the interest.'

'Interest, interest? Why, there is no interest!'

'No interest?'

'No, didn't you say, when you brought your money, that you had brought it to place in my hands?'

'I did, Mr Wood.'

'Did you say you wanted it put out at interest?'

'No, I did not think it necessary, Mr Wood.'

'It has not been any use to me. I have been afraid to use it or invest it; I have kept it wrapped up in my drawer. If you had wanted interest you should have said so. No, no, I be very sorry you did not mention it; I have kept your money safe, and it ain't made any interest.'

There are many other anecdotes about Jemmy Wood in a book published by C.H. Savory in 1882.

In spite of his ways, he had no fear of losing customers to his competitors. He was known far and wide for looking after his own money—and was judged equally worthy of looking after other people's.

Wood saw off several other banks in Gloucester: Turner and Co., Stephens and Co., Fendall and Co. and Jeff and Evans, bankers in Northgate Street, who were heavily committed to the financially crippling attempt to tunnel under the Severn at Bullo, near Newnham, in around 1817.

In 1808 he became a member of the city corporation, in 1811 a sheriff, and in 1820 alderman. He had been known to say that day he would 'do something for poor Gloucester,' but he would not subscribe charities.

Most of his time was spent in his bank, or on visiting his tenanted farms. Some mornings he would go down to the Quay to glean coal dropped from the barges, while on Sunday he went for long walks in the country and gathered wool he found clinging to fences. He never entertained anyone; apart from his assistants, Osborne and John Surman, his only friends were Sir Matthew Wood and a Gloucester solicitor named John Chadborn.

Sir Matthew, who was no relation, was a Member of Parliament and lived in London. Chadborn conducted business for Jemmy, who thought he was kind and disinterested, as he appeared to make no charge except for out-of-pocket expenses, but unknown to him was keeping a detailed note of all his services.

Wood, the last of his family—he had two sisters who died at an early age—remained a bachelor. He once confessed that he courted a young lady at Dymock, and loved her dearly, but she had jilted him and married another: 'It was too bad of her, too bad ... I never could trust another.'

Apparently he did pay attention to another: a Miss Anne Weekes of London Road, who was the owner of a ferocious dog. In the end Jemmy said she would have to choose between him and the dog, and she chose the dog.

To the day of his death, April 20, 1836, he kept his bank and shop. By then his fortune amounted to a million pounds, plus a vast income from his investments, and his fame had spread to almost everywhere in Britain.

As he lay dying his executors and beneficiaries, Sir Matthew Wood, John Chadborn, Jacob Osborne and John Surman, gathered at his bed. There was even a scandalous rumour that they aided the old man's exit from the world, and there was certainly scandalous behaviour and levity at his funeral.

This was nothing, compared with the behaviour of his four executors. During his life Jemmy Wood had been quick to give advice to others about making wills, but had been reluctant to put his words into practice. Just before the funeral procession started a James Wood of Islington introduced himself as a relative and asked to be allowed to attend the funeral. Permission was refused, but he went to the church all the same, and afterwards entered a caveat against probate being granted to the executors.

The will presented by the executors, which nominated them sole beneficiaries, consisted of two papers stuck together. Singularly each was deficient, but jointly they explained each other. One paper was dated October 2 1834, but bore no signatures of witnesses; the other was undated, but did bear signatures.

The next development was a letter in *The Times* from a Thomas Wood junior of Stamford Street, claiming that his father, Thomas Wood, Thomas Helps of London and James Helps of Gloucester were among the next of kin. Then a charred codicil dated July 1835 was sent anonymously to Thomas Helps, naming a number of new legatees, who benefited to the tune of some £150,000; there was also £60,000 credited to Gloucester Corporation, in addition to £140,000 said to have been left to the corporation by a former codicil. The envelope which contained it bore a London postmark and a pencilled note said: 'The enclosed is a paper saved out of many burnt by parties I could hang; they pretend it is not Wood's hand, may swear to it; they want to swindle me; let the rest know.' It was unsigned.

Rewards were offered to the sender if he would come forward: Chadborn offered a hundred guineas; the corporation offered £2,000; Thomas Helps offered £10,000; the corporation offered a further £5,000. But none owned up.

Then the legal wrangle began. Accusations and insults were cast, and the executors were charged with false pretences, which they countered with charges of forgery. Litigation continued for five years; in 1839 the Prerogative Court of Canterbury rejected the codicil as spurious but in 1841, after an appeal, the Privy Council reversed the decision and found for the legatees. The executors won probate, and after paying out the legacies and costs kept some £400,000 for themselves.

The corporation was disbarred from its inheritance on the grounds that it was a trustee, and had no specific information on how it was to employ the money. It pursued the matter further, rejected an out-of-court offer of £50,000 from the executors, lost its case when it did reach court—and was forced to levy a rate in order to pay its legal costs.

At some stage in the legal proceedings the executors produced forty witnesses to try to prove that Jemmy Wood hated the corporation as much as he did charities. In the course of this the executors were forced to confess that they had stuck the two wills together fraudulently, and Chadborn admitted to burning some of Jemmy's papers. Doubtless all of them took unpardonable liberties with their trust, and it weighed heavily on some of them. Chadborn hanged himself, allegedly because he was so upset by the judge's severe remarks.

The National Provincial Bank established a Gloucester branch in King Street in 1833 or soon afterwards, and three or four years later it bought Wood's premises, rebuilt them and stayed there until it moved into new premises in Eastgate in 1889. By this time the sober world of institutional banking had long overtaken the maverick individuals, and the remarkable Jemmy Wood was nothing more than a folk memory in the city's consciousness.

# Foremost reasons to drop the pilot

June 1986

A well-tended and productive vegetable garden is both laudable and a source of satisfaction. I admire those hardy people who work in their gardens early in the year, undeterred by biting east winds or lack of sun, but my admiration does not extend to emulation. In late March or early April I make a reluctant start in the garden, consoling or excusing my tardy beginning with the old belief that you should not plant the garden until the soil is warm enough to sit upon with bared bottom. I have never actually tried this experiment; I just wait until the sun is warm enough to cast a glow on my back.

By May my reluctance has turned to enthusiasm fed by the gathering strength of the sun and by the first results of my earlier endeavours. Plying a hoe between the rows of emerging seedlings is a pleasant job; the time to hoe is before there are any weeds, so with a bit of luck my garden looks presentable by early June. Then June and July bring silage and haymaking on the farm and my absence from the garden and in this time, of course, more things grow in it than the ones I have planted.

Faced with a jungle of weeds, I am a reluctant gardener again by August—but by September I feel a renewed enthusiasm. In October I think I shall perhaps plant some early potatoes. It may seem a strange time, but the best and most hardy earlies are the ones which come up by chance, from potatoes missed and left in the ground over winter. By planting in autumn I can even steal a march over all those hardy gardeners.

This year I have deserted *Arran Pilot* for *Foremost*. I feel a bit guilty about abandoning an old favourite, but *Foremost* have a better flavour—and the palate, I believe, should be the chief guide to our choice of varieties. Not the only one; there are other important considerations such as immunity from disease, ease of growing and cropping capacity, all of which *Foremost* potatoes also possess. So, by the way, do *Scarlet Emperor* runner beans. Until I grew them last year I never really had a taste for runners.

Another change has been forced on me this year. The frosts and cold winds of February destroyed the autumn-sown broad beans, and I have had to replant with spring beans. This, I am afraid, will upset my usual plan of leeks following broad beans, as it is doubtful if the springs will be cleared in time for leeks.

It was broad beans and leeks which started my no-digging method in some parts of the garden a few years ago. It seemed a pity to dig and lose both moisture and the benefit of the bean roots with their nodules of nitrogen, so I hoed the ground to make a shallow tilth, spread some compost and just put the leeks in with a dibber. I thought there might be an increased problem with weeds, but there were far fewer than usual. Succeeding crops also caused no headaches. Sprouts, onions, carrots and parsnips have had straight, unforked roots.

This no-digging method would not be successful without compost. It provides fertility and makes the soil friable as well as increasing the earthworm population. I feel the worms are doing the digging instead of me. Not that it is entirely without labour. It takes a lot of waste vegetable material to make a small amount of compost, and the compost has to be made properly. The heaps have to be built fairly quickly to build up heat to destroy weed seeds and promote decomposition, but at least you have an added incentive to collect weeds, trimmings, lawn mowings and so on.

Compost is improved if it is made from a great variety of materials. Lawn mowings are excellent, especially for building up the temperature, but too many at once are disastrous. Of course I am lucky because I can use manure as an ingredient; as well as providing much-needed bulk, this helps to activate the heap, but at times I have used too much, with unwelcome results. Like tightrope walkers, compost-makers know that balance is all.

# The family who clung to old ways

July 1985

The death of Miss Alex Dowdeswell of Wick Court at Arlingham, by the Severn in the Vale, meant the end of an old-established, old-fashioned and respected Gloucestershire farming family. For many years it was noted for its Gloucestershire cattle, and it was the dispersal sale of the herd in October, 1972, that led to a revival of interest in this rare breed and the reformation of the Gloucestershire Cattle Society.

There are few, if any, such farming families left in the county, families that have clung to the old ways and kept the traditional, long obsolete tools and implements of farming. Like Miss Alex, neither her brother Robert nor her sister Miss Ella, both of whom died some years ago, had married, so there were no children or grandchildren who might have brought new ways to their family. So the cattle remained until the sisters became too old to attend to them; and the old implements stayed in the sheds long after they were of any practical value, passing through the stages of use to uselessness, from rubbish to collectors' valued pieces.

At the recent sale that followed Miss Alex's death, there they were, some lined up in a field, others still in sheds, all in various stages of decrepitude and bearing the auctioneer's ticket and a lot number. There was old dairy equipment too—buckets, three-legged stools, butter-churns, cheese rings, a horse-drawn milk float.

In the field and in the sheds I saw both horse—and tractor-drawn ploughs, cultivators, a corn-binder, a wagon, horse-drawn corn and root drills and a fertiliser distributor; a dung cart, mowing machines, hay-turners, a hay-rake and harrows, all made in the days of horse power. And I remember using implements identical to almost every one of these, not as curiosities but as everyday objects.

When I used them I never realised that I should ever see them all discarded, or as mere museum pieces, yet I am not yet quite sixty. What a change in one man's lifetime. Gone, gone, the old attitudes, the old ideas of husbandry, the Gloucestershire and Shorthorn cattle of the Vale, the horses and their implements, even the men of the land and their ancient skills, all in half a century or less.

As I stood in that field near the willows, gazing at those old implements, it made me feel old. Yet I am glad I am old; not so old as some who stood beside me, but old enough, and glad to be old enough to remember those days, the tangible remnants of which I saw before me. I am glad to have carted dung with a horse, to have tipped it into heaps and spread it with a fork, to have horse-hoed roots and singled them with a hoe.

I feel privileged to have forked sheaves, to have helped build corn and hay ricks, to have heard the rumble of wagons, the rustle of sheaves of corn, the jingle of chains, to have smelt the sweat of horses and all those other scents now absent from the farming life. Above all, I feel privileged to have known and to have worked with men who could hoe, plough, thatch and hedge and to have enjoyed their company, shared their talk and jokes; and to have sat under the hedge, or by the side of a rick, eating cold fat bacon and drinking sharp cider.

I was looking at some old farming photographs the other day and there were such men, sitting down and eating their 'bait'; all of them wearing stout hobnailed boots. Once such boots could be found in quantity in every market town and many village shops, heavy boots so stiff until weeks of dubbin and wear made them pliant. All farmworkers wore them, and perhaps 'yorks' as well, straps or string round their trousers just below the knee. Except on the wettest days in winter, hardly anyone wore gumboots. Like their boots, these men had solidarity; and with families such as the Dowdeswells, they shared a sense of duty to the land.

Times have changed; there have been improvements, too, but the farming world is poorer without the horses and the men who worked with them and with hand tools, and without the Dowdeswells of this world. Deny it if you will, but these people had dignity and they had style, and we should honour their memory.

# A toppin' bad season of it

## July 1986

In the days when we had a local fruit market, a meeting used to be held some time in July to discuss the prospects and the marketing of the plum crop. Or rather two meetings would be held, an official one in a large room at the pub adjacent to the market, and an unofficial one in the bar where we gathered for a drink beforehand.

Invariably the real meeting had to wait. Half an hour after its advertised starting time, and in spite of the entreaties of those responsible for its organisation, many plum growers were loth to move from the bar, especially the ones in full spate with complaints about the previous year's marketing arrangements.

There was one old grizzled man, too, who would not be hurried on principle. 'They ain't got no right to order us chaps about,' he proclaimed. 'Don't you allow yourselves to be ordered about,' and there he would sit in a corner of the bar, resolute and scornful of those who allowed themselves to be ushered into the other room.

Then, after a desperate plea from the auctioneer, he would struggle to his feet and glare at his dwindled number of loyal companions: 'Come on, you chaps, it's high time you was at the meeting.' Turning to the auctioneer he would say: 'I've been tellin' 'em, Sir, as they oughta be in the meetin', but they 'ouldn't shift.'

Once all the company was assembled in the big room, the auctioneer addressed the meeting. 'Quite right, Sir,' boomed the grizzled man almost immediately, 'it's real good on you to come.' The auctioneer thanked him, and in spite of several further interruptions, managed to continue with the business.

'What sort of a crop do you think you have?' he eventually asked the plum growers. 'Fair,' said one. 'Fair to middlin',' said another. 'Goodish,' said a younger man.

'What!' roared the grizzled man, clambering to his feet and trembling with amazement and indignation, 'Goodish, did you say? What do you mean? Do you know what you be on about? Fair, you over there said, and fair to middlin'. What kind of talk is that?'

'They don't know what they be talkin' about, Sir. Oh, they might have a feow plums on a feow trees, but that ain't what you meant, Sir was it? You meant a real assessment over all, an' I tell 'ee 'tis a patchy ol' crop this year an' no mistake. No, Sir, I'd go further, an' say 'tis a poorish ol' crop, a poorish ol' crop Sir, an' I d'know. I bin about a long time an' I d'know, an' I got eyes in me yud if some on 'em an't, an' I get about, an' I d'know.'

Usually no one liked to venture any further opinion after this outburst. The grizzled man would sit down, quietly triumphant. The auctioneer would smile and say: 'A patchy crop, fair in places, poor in others.' 'Mostly poor, mark my words, come pickin' time an' you'll see,' muttered the grizzled one. In July most plum growers were reluctant to admit that they had a good crop, just as salmon fishers shrink from declaring that they are having a good season. 'Now to marketing arrangements this year ...' began the auctioneer.

'Ah, I'm glad you've mentioned that. An' I hope you'll make a better job on't this year. Last year you made a right mess on't to my way of thinkin'.'

'We could make a better job of it if I could explain without interruptions.'

'Quite right, Sir. How do you explain, an' how can we understand, if there's constant interruptions?'

'But it's you who's interrupting.'

'Me? I ain't interruptin', I'm tryin' to help you, that's what I'm doin'. You'll need all the help you can get, Sir. You'll have a difficult job this year with all these plums about.'

Tactfully the auctioneer took no notice of this last remark, apart from a nod and a wink to his colleagues, unnoticed by the grizzled man.

'Non-returnables this year,' said the auctioneer, and a colleague held up a box for our inspection. The non-returnable box was passed round and examined critically. 'No good at all for puttin' food in for calves,' announced the grizzled man.

'They are for plums, not for calf feed.'

'Ah, but the wooden 'uns was toppers for calf food.'

The rest of us had a chance to speak; but not for long ...

'Well, I still don' like 'em. An' they ain't strong enough for plums. When us gets 'em back next year they'll all be bust.'

'You won't get them back, they're non-returnables.'

'Well I'll get 'em back meself an' use 'em again.'

Eventually the meeting came to a close, the auctioneer thanked us for coming, and hoped we would have a reasonable crop, in spite of the dismal forecast. Slowly the grizzled man rose to his feet, thumped the floor with his stick and boomed: 'An' thank you fer comin' an' explainin' everythin'. You made a toppin' job on't last year, an' I'm sure as you'll make a toppin' job on't this year.'

# It isn't the noisiest pig …

July 1981

In my tribute to H.J. Massingham in May, I wrote that he wanted 'not a quiet countryside but a busy one.' By one of life's mysteries, 'busy' became 'noisy,' which altered the sense completely. Noisy is not synonymous with busy. As the old saying has it: 'It isn't the noisiest pig that eats the most food.' And Massingham of all people certainly did not want a noisy countryside.

But a noisy countryside we have, especially at certain times of year. Who was it who said he would have to go into town for a while for a bit of peace and quiet? Noise can be one of the worst forms of pollution, spreading so widely, but with mechanised agriculture it is unavoidable, with only its degree and duration the point at issue.

Tractors, balers and combines are tolerable; forage harvesters, chain saws and irrigation plants are the worst offenders and zooming low-flying aircraft. They make me jump, but surprisingly cattle are rarely startled by them. Aircraft are defence, though, not agriculture; even so, I sometimes wish they would go and defend somebody else, and wonder if the time will come when our obsession with defence will leave us with precious little worth defending.

Back to earth, we are glad enough to make use of forage harvesters and chain saws, so I am not complaining about them as such; one never does complain about the things one finds personally useful. But there comes a point where their use is objectionable, unnecessary and unsociable.

There is none of the element of desperate haste with sawing wood or silage-making that there is with haymaking or harvesting, and there is certainly no good reason why they should be done on Sundays or on weekdays after 8:00 p.m.

Irrigation plants whine continuously all day long, weekdays and Sundays and often far into the night, especially—if you can remember back so far—during periods of drought. Absurdly, perhaps, I believe their use should be banned during severe drought, when the streams and rivers are at their lowest. But even less defensible and even more disturbing and maddening is clay pigeon shooting on Sundays.

With all this strident noise assailing our ears we are denied the gentle, enjoyable sounds, the song of birds, the rustle of leaves. And intensive mechanisation has deprived us of other noises, including pleasant mechanical ones: the swish of a scythe cutting grass, the sound of stone on blade as it was sharpened, even the noise of the old manual lawnmower.

The sound of the horse-drawn mowing machine drifting across the meadows on a summer morning, the click-click-click of the gears as it was reversed on a corner; men's voices shouting commands to their horses. The clackety-clack-clack of the old binders in the cornfield—rather a jolly sound: 'clackety-clack-clack, we're getting on with it, clackety-clack.'

The rumble of wagon wheels, the rustle of hay or sheaves; the clink of chains, the sound of iron-shod hoofs on cobbles as horses returned from work in the fields. The ting of milk hitting the bottom of a pail, changing to a fuzz-fuzz as it filled it, in the days when we sat on three-legged stools to milk cows, listening to the comfortable munching of hay. The sound of steam engines, the long-drawn-out hiss as the train reached the station, a huge mechanical sigh of relief saying: 'I'm here at last.'

How dear to me were those sounds when I heard them, and how precious they are now I hear them only in memory. I hug my good fortune for having heard them and for being able to hear them so clearly now. Will my children, or my children's children, feel the same about the noise of the forage harvester? Perhaps they will, if they are cursed with even more clamorous machines.

# Plum perky

August 1979

Our earliest plums produced in any quantity are Czars; the Czar of Russia was visiting this country when the strain was introduced, and it was named in his honour. It was also called the seaside plum—perhaps it still is—because it was ready in time for the Bank Holiday trade on the first Monday of August.

Changing the date of the holiday to the last Monday of the month has played old Harry with our maincrop plum, the Blaisdon Red, named after the village in which it was first discovered. Most of these plums, produced in great quantity in this district, go for processing. But the factories close down for some days around the holiday, which means that picking has to stop on the preceding Wednesday or Thursday, and cannot start again until the holiday itself, at the time when the bulk of the plums must be picked.

To make matters worse, most plums are picked by casual workers who try to arrange a week's holiday to coincide with plum picking. It is a gamble. Do they take the spell preceding the holiday, in the hope that the plums will be ready, but accept that they will miss part of a week's picking? Or do they choose the following week, hoping that the plums will be late? In any case, those vital weekend days are lost.

The pickers don't like it, the growers don't like it, the merchants don't like it—and eventually the factories don't like it, either. The plums for canning need to be firm and only just coloured, those for jam riper, but all too often the former arrive too ripe, the latter too green; and a day or two's rain after the holiday, when the plums are too ripe usually means a fall and a considerable loss to the grower.

Victoria plums, the favourite with the public, also clash with the Blaisdons—as, so often, does the corn harvest, too. Last year there were heavy crops of most plums—and two heavy lots of Victorias, in fact, the English crop and the one imported from Italy.

This year we have had a good blossom, although the weather was atrocious while it was out. At the time of writing it's too early to say if we shall have a good crop, for as an old neighbouring farmer used to say: 'Blow ain't plums!' All I know for certain is that if there aren't any plums, everybody will want them, and if they are abundant, there will be no demand.

In the expectation of a good crop, perhaps I can stimulate interest with a true story. We used to have an old farmer who grew a considerable number of plums, and he was over eighty when we asked him the secret of his remarkable vitality, one night, after several drinks in The Junction.

Perhaps I should explain that besides his plum orchards he had a bicycle; and a girl friend some six miles distant, whom he visited frequently.

'The secret of my vitality?' he echoed. 'Well, 'tis no secret really, 'tis quite simple an' what's more I'll tell you.'

He paused, pleased and perky, as we waited for an answer.

"Durin' the season I eat as many Blaisdons as possible, an' their juice an' acid an' goodness stores up in me body an' keeps me goin' the whole year long. 'Tis true, I tell you. An' I be the livin' lively proof on't!"

# Days of Dick Turpin
# and the nogmen ...

Oats were the earliest corn to be cut in the days of the binder, and they remained in stooks until the church bells had rung for three Sundays. In the last war, however, they had to forgo this pleasure, for bells were to ring only in the event of an invasion.

The first job of harvest was to hand-cut a passage round the field for the binder. Using hooks and crooks, my father and one or two other men made inroads into the corn, leaving it in neat sheaves for me to tie with a dozen straws from each. The 'War Ag' men used to drive straight into the standing corn, but my father called that a lazy and slovenly way. Not that we saved very much corn with our methods, for the hand-cut and bound stooks stacked against the hedge soon suffered the depredations of birds.

The binder would be dragged from its hibernation, cleaned of the several oddments it had accumulated during the winter, and oiled. Then the annual argument would begin. There were three canvases to be fixed, which took the corn from the knife to the knotter. Each was slightly different in length, and each had to be put in the right way round.

Everyone had a different idea about which canvas went where, and which way round. This of course led to several computations, frequent changes of mind and contradictions, making an apparently simple matter rather complicated. The epithet 'nogman' was applied to all except my father, whose ideas about how the canvases should be fixed the rest of us secretly hoped would be wrong in every way and direction.

Eventually, by trial and error, all the canvases would be fixed correctly. 'Before we take them off at the end of harvest,' my father used to say, we should mark them. It would save this bother every year.' We never did mark them, and we had that bother every year. Perhaps we did not mark them because 'the bother' was all part of the ritual of harvest, and we rather enjoyed it.

My father had a tractor before the war, but I remember when the binder was pulled by horses. It was a hard slogging job for horses, as many other jobs were. The tractor was a tremendous boon to farmer and farmworker and we welcomed its arrival. I dare say the horses did, too, swishing their tails in the shade of a tree. Horses on mainly pasture farms, although they did not face as much hard work as those on arable ones, probably fared the worse by being put to spasmodic and sudden hard work to which they were unaccustomed, and working with other horses with which they were unfamiliar.

One of my father's horses, Lloyd George by name, although nice enough in other ways, was lazy. He would lag behind and let poor old willing Dick Turpin do most of the work. When the pair of them were hitched to the binder, I have seen my father walking beside Lloyd and flicking him with a swishy stick to goad him into pulling his weight.

It is easy to become nostalgic about horses, even if they were not all like Dick. There were horses that were determined not to be caught in the field, horses that were recalcitrant or a struggle to harness, ones that would feign lameness, and clumsy, awkward horses.

There were horses that would not go, horses that would not stop, wild ones and nervous ones. Nasty, ill-tempered ones whose only desire seemed to be to kick. But there were also the patient, trustworthy ones who would work with a will, like old Dick. One can regret the passing of horses, but how we welcomed the tractors.

With the plough-up campaign well under way during the war, almost every farmer wanted a tractor, but had to wait his turn. My grandfather and uncle were exceptions, but even they eventually bought one. My grandfather never shared the general enthusiasm for tractors; neither did my uncle when theirs first arrived, but within a few weeks he was saying: 'You can do more after tea with a tractor than you can do all day with a pair of horses.'

Horses had ruled the farm, and the length of the working day with such jobs as ploughing, but tractors did not tire. For a little while man was master. Soon, he found he was working to the tractor's dictates, and today the patterns of farming and of the fields are set solely by machinery.

Tractors and other machinery are so much quicker than horses. Yet while few farmers felt compelled to work on Sundays at hay or harvest in the days of the horse, now, with tractors, balers and combines, almost all of them do.

# Gardening

August 1977

If there is one human occupation that has general approval, it is gardening. Gardeners and non-gardeners alike, when we learn that so-and-so is fond of gardening, we warm towards him. Even if it is someone we never really liked, once we know he is a gardener, we think 'there must be some good in the man,' and picture him, his face not so hard as we had supposed, lovingly tending his roses or cabbages.

These days it is more likely to be cabbages, with the high price of vegetables. I understand that in Bristol alone there are several hundreds waiting for allotments. But there is more to it than saving—although the saving is considerable, when you consider that, with taxation, the average worker has to earn a pound in order to buy sixty-five pence worth of vegetables. More to it, too, than the freshness, the improved taste of home grown vegetables, and the sheer satisfaction of eating one's own produce.

Even greater than these is the joy of gardening. Not an alloyed joy. Happiness is a matter of light and shade, and were it not so, we would call it boredom. There are backaches and disappointments, crop failures, weeds, pests and disease, the vagaries of the weather. These perhaps, make the townsman understand the farmer's problems better, to begin to know why he has a reputation for grumbling. As the old rhyme reminds us, though, the farmer and the gardener are not always in agreement:

> The farmer and the gardener are both at church again,
> The one to pray for sunshine, the other to pray for rain.

August is the time of flower and vegetable shows, which bring out the best and the worst in gardeners. The best is displayed on the trestle tables. The worst is displayed when their best is not considered the best by the judges.

There are marquees where there are bitter looks, barely suppressed hostility. Not in my parish of course. Well, there was one occasion. A competitor was disqualified; it was said that the judges had visited his garden, the onions he had exhibited would not fit into the depressions left in his onion bed, and there were other discrepancies. He set up a stall outside the showground, displaying a placard proclaiming his innocence. His mother swore to sell her house to provide the money to clear his name. But it was all a long time ago.

# When mud gave way to Meadowsweet

## August 1983

This spring our lanes were lined with lace, the lace of cow parsley in bloom, a more glorious bloom than I can ever remember seeing before; miles of luxuriant white lace—until the council's machine came and haggled it off, leaving untidy, bent and crushed stalks to wither and litter the roads.

In the one lane that escaped the depredations of the machine, the cow parsley was allowed to flourish—not to mention that great ungainly giant the hogweed, cranesbill, buttercups, herb robert, red, white and yellow clover, vetches, an occasional bright red poppy, meadowsweet, fluffy and feathery grasses and arching tall fescue which dropped into the road.

Rather a nuisance, this tall fescue, especially when dampened with the rain, but better this than losing the wild flowers: the white moon daisies and pink brown sorrel which, hot or cold, wet or dry, proclaim the reality of summer. Ragged robin in damp places, scarlet pimpernel in cultivated ground ...

Weeds, men call them, these and other plants of verge and hedgerow, bank and ditch, field and woodland, but:

> What would the world be, once bereft
> Of wet and wilderness? Let them be left
> O let them be left, wilderness and wet;
> Long live the weeds and the wilderness yet.

The glory of elderflower this year rivalled that of the cow parsley, such large creamy-white heads, and in such profusion. Pink and white dog roses waving in the hedgerows, more delicate and carefree than our cultivated varieties. Small white flowers of the blackberry, the honeysuckle casting its glow and scent: all these delights and more along a short stretch of road. There was the stink of pig slurry a neighbour had been spreading, too, but where the honeysuckle grew its scent overcame the stench.

Several times a day I travelled this road as I hauled the cousins of those wayside weeds for silage and hay. Loads of tails I had: cats-tails, fox-tails, dogs-tails as well as cocksfoot and birds-foot. I also had ryegrass, a dull name for a dull but useful crop; red and white clover; fescues and other grasses.

The cursed rain that made us despair in April and May produced an abundance of grass; this year where we gathered three loads there was scarcely one last year. We can do nothing about the weather, and what a good job we cannot, for left to itself it eventually puts everything right. If man could control it what chaos there would be. Mind you, during these past few years the weather has been very tiresome. The rain has never come when we have wanted it, and at other times it has seemed as if it would never stop.

We and our animals squelched in mud this spring, and more and yet more rain, day after day, with the sun not daring to show its face. Then at last it shone as of old, and our world was transformed. Everybody smiled, it was safe to speak to farmers again without running the risk of a glower and surly reply, the cattle grew sleek and shone in the sunlight, and the land could carry a tractor again. Within a week the earth became hard where before it had been a morass; and farmers began to say to each other: 'We could do with a drop of rain.'

# Writers who missed the rhythm of life

August 1984

In summertime I often envy those who do not work on the land, because they are the ones who can appreciate the countryside in summer; we landworkers are too busy, what with silage-making, haymaking and harvest, and like W.H. Davies I find myself wondering:

> What is this life if, full of care,
> We have no time to stand and stare?

In the hedgerows the wild roses nod an invitation to stop a while to admire their beauty. In the cornfield the demure green ears of barley sweep and bow in the gentlest of breezes; bees are busy on the white clover of the pastures, and campion, vetch and stately hogweed decorate the roadside until the council mower comes along.

There is sunlight and shadow, too; I could certainly stand and stare at sunlight and shadow, though they never stand still, and how cool and inviting is the stream-side shade of the willow and alder when the heat is shimmering over the hay-field in which we toil and sweat. Again, how I wish to stand and stare at the clear, hard light and at our red cows all sleek and ashine.

Or so I think when there is no chance of doing so, when there are bales to be gathered or corn is to be harvested. When the chance does come, these sights lose something of their allure, partly because the moment you become an observer, you are no longer a part of the picture; this, in my view, is the failing of so many country writers.

Most of them saw, but were not part of the scene; they witnessed the toil of the fields, but knew nothing of the rhythm of that toil. Hardy only observed, and often drew the wrong conclusions from what he saw. And though the son of a small farmer, Richard Jefferies, too, was purely an onlooker. Had he been disposed or able to have helped his father—who must often have needed an extra hand—he would not have made such disparaging remarks about land-workers or their labours. True, the farm-worker's conditions were hard, and his rewards poor, but Jefferies had no understanding of the pace and pattern of his work.

Machinery has taken much of the toil out of farm life since the days of Hardy and Jefferies, but it has taken away the rhythm, too; and in taking away the men, it has taken away the companionship of the fields.

Machinery does sometimes give us time to stand and stare at busy times. We stand and stare in stores while waiting for some vital spare part, which often enough is not forthcoming. In these days of computers, speed and efficiency, we are sometimes forced to wait for days because a small piece of metal is not available; occasionally, indeed, it is there in stock—but it is still not available if the computer says no.

Such delays would not have been tolerated, and were not even experienced, in the days of our grandfathers. No doubt they never talked of efficiency, but they went on, slowly and steadily, day after day.

Harvest with binder, sheaves, wagons and stacks encapsulated the rhythm that was once a part of farmwork. Rhythm and pattern; round and round went the binder, throwing out sheaves; round and round went men in the opposite direction, picking up sheaves, one under each arm. Men met and sheaves met and were stooked, and so on, in a kind of ritual, a rural dance to the sound of the binder's 'clack-clack.'

More rhythm and pattern to pitching and loading the wagons, to unloading and stacking. Each sheaf placed correctly, and in time with the next man. On load and stack each sheaf was positioned not only for building, but for the subsequent unbuilding, too; each formed part of an intricate pattern for utility which also resulted in beauty, as it does in all good craftsmanship.

The tools the men used became part of them, an extension of their arms. The apparently clumsy scythe was a thing of beauty and grace in the hands of a skilled man.

Energy was conserved because of necessity; but today it is a different story, with men working not to a natural tempo but to the demands of that tyrannical new master, machinery.

# High and low,
# we're all scratching a living ...

August 1981

Everyone was agreed; no-one could remember a spring like it, with snow, frost and biting winds at the end of April. Cows that had been turned out to grass were hurriedly brought inside again, and their milk yields fell, never to recover.

May, too, was cold and wet and cows, outside again, churned the pastures into a morass, their udders plastered with mud, as they had been when grazing kale in a wet November. A lot of spring corn had been left unplanted, and that that had been looked sick. With more rain, the barley was planted late; barley in dust, as the adage has it, though this year it was barley in mud. Rain, and more rain, and more mud; those who tried to make silage made ruts in the fields.

It was 'the damndest, wurstest spring' in living memory—in fact no spring at all. Farmers gazed disconsolately at their sodden fields, crops and stock and predicted ruin. The only consolation was that everyone was in the same predicament, but that was not much consolation at all as the rain came down day after day. It had been like it a hundred years ago, and there had been no summer then, either. We would have no summer this year, it stood to reason. The tag about wet to dry and dry to wet paying a debt; it was not going to, not this year.

Only the little man in the pub offered a ray of hope and sunshine; by the twenty-second of June the fine weather would arrive, he said. He had, to be sure, been right about winter in April, but as June came in wet we lost confidence even in his predictions.

Then it stopped raining all day and every day; some days it did not rain at all, though as our hopes revived, down it came again. A few fine days and the mowing machines were busy; silage-making began; we cut some grass for hay and baled it on a hot sunny day, so hot that we drooped like cut flowers without water. I looked at the calendar and saw it was the twenty-second of June. We cut more grass, but would the fine weather last? No-one dared go and ask the little man; when ignorance is bliss ...

A kestrel hovered all day over the mowing field, waiting to pounce. A crow swooped down and carried away a mouse. The kestrel seemed to have the more difficult job, hovering in the air for hours on end. A couple of crows turned nasty with the kestrel. The wild roses were in bloom in the hedgerow, pink and white ones, larger than usual and with beautiful golden centres. The elderflower blossomed, and the meadowsweet; could be summer after all.

In the field we cleared of grass for silage I saw earth being heaved up by a mole. A mole can go for only a few hours without food; it has to keep burrowing and heaving away in search of it, but not so deeply this June. The moist ground had brought the worms almost to the surface, but it still seems a lot of digging, to me.

It is a hard way to make a living, burrowing all day long and most of the night, though I suppose a mole hardly knows if it is day or night, anyway. Not that things can be much easier for the kestrel; up there in the sky, vigilant on the wing, that must be a risky way of scraping an existence, too.

# What has chaos got to offer?

August 1980

It was a sequestered hamlet; a score-and-a-half houses, a post office-cum-grocer's shop, a pub and a railway station with an adequate train service.

A few council houses, pleasantly situated and blending well with the rest, were added to fulfil a local need, and about a dozen more houses were built in as many years. No-one objected, and they did little to alter the essential character of the community—although, of course, it was inevitably altering, as more people were forced to seek employment away from the district.

Electricity had already arrived, and mains water soon followed. Both were boons, but the latter brought drainage problems, the primitive system being unable to cope with all the new bathrooms, water closets, and the rest; in a hot summer you could smell the problem.

But the hamlet was nice and peaceful, and looked like remaining so, especially as the planners were reluctant to allow any more houses. At that time, with an excellent train service, a good case could have been made for more; but the authorities said no.

The train service was stopped, and the station demolished—and after that, more houses *were* erected. One supposes that planners discuss the design of houses and even visit the sites; but after seeing the results, it is often hard to believe.

These are houses which are admirable for town or suburb, but totally out of place in the country. It is a pity that the townsman who comes to live in the country has to bring the town with him—and it is a pity that it is apparently beyond the wit and skill of planners and architects to design houses to fit into rural villages. Expense is usually given as the reason, but it is only an excuse, and a poor one at that.

Eventually a main sewer was laid and a few more houses were allowed. But when a farmer wanted to build a house for his son who worked on a neighbouring farm, permission was refused; the sewer could not possibly take another house.

After that there was a lull—but then more homes began to go up; not for farmers' sons or other local workers, and not even, in fact, for people who wanted to live there specifically. As soon as the new houses neared completion, sale notices appeared; this, I am told, is what constitutes a demand for houses in the hamlet, although the number of sale notices and the time that they are up could lead one to believe it was something rather different. Meanwhile, in another hamlet where there is a genuine demand for two or three houses, permission is refused.

The time came when the local people and the newcomers thought enough was enough, and turned out in force to a parish council meeting to object to further development. None of the parish council lived in the hamlet, and with the exception of one member, the body was unsympathetic to the case presented by the inhabitants.

'Only three more houses,' it was argued, 'surely you can't object to just three more.' The inhabitants were wiser, and knew it would be three more and three more, then six more, and so on. Too late, the council saw the warning light—and now its objections, too, are proving futile.

So it goes on; no-one knows where it will stop. Planting houses pays better than planting corn; a garden is dug and another house is planted. If there were less coyness from the planners, if there was a genuine demand, if there was local employment, if the trains were still stopping, if there were other facilities, perhaps it would make more sense. There are a dozen ifs.

Rumours are rife—fifty houses in this field, a hundred in that? Has that old neighbour been seduced by Mammon, and is he going to try to plant a few in his garden? How are the children going to fare when they leave school? Will miles of country lanes be widened? Will the hamlet be swallowed up by suburbia?

And meanwhile, how is the sewer coping? And if this is planning, what has chaos got to offer?

# Slippery path to anarchy ...

August 1985

In the large parish in which I live, several of the most pleasant public footpaths were virtually unusable for many years. Their footbridges over streams or deep ditches had fallen into decay, collapsed into stream or ditch and become lost with the water which has flowed beneath them, and with the passing years.

This year, though, Gloucestershire County Council has put up new bridges, reviving old paths and bringing old delights to a new public. For delights there will certainly be for those who care to tread them through the fields and wild flowers, and along the banks of the willow and alder-lined streams. A peaceful pursuit, this, and a rewarding one, a refreshment for the spirit, an antidote to the modern world's mad rush. I believe the council is to be commended on its work.

We who live in the country are fortunate to have the fields around us. How many people who live in the town can walk through such surroundings when their day's work is done, as the poorest countryman could once do?

Walking footpaths has become a popular pursuit, and if townspeople are using them it is an opportunity for us all to build bridges between the town and the country. It is a gap which needs bridging, since we are in danger of becoming two nations, town and country—and a nation divided is no sort of nation.

Not every landowner shares this opinion. Neither do they share my opinion that footpath-walkers, by the very nature of their occupation, are peaceful and law-abiding citizens, in the main. Some would stop people from walking along the public footpaths over their land, while even the large, powerful landowners of long ago tolerated it, though I expect they could always have prevented it, law or no law. This only goes to show that some of those old overlords were not so bad as they have been presented, and that some of today's petty landowners may not be so good. Of course, the peasant footpath-walkers of old respected crops and stock and property—but as I have said, I have reason to believe that most of the present users would, too.

Be that as it may, the re-erection of footbridges has not met with general acclaim. I do not know whether what I have heard is strictly accurate, and I certainly hope that the forecasts are not, but it appears that a reign of terror is about to descend on us—all because of a few new footbridges and some old footpaths.

Rumour has it that we are likely to have arson, burglary and vandalism. Gangs of terrorists and savage dogs will with wild halloo roam the paths, committing all sort of damage. Sheep will be killed, and cows will abort at the sight of them.

I am well aware that there are vandals and burglars, but I never realised that among their numbers were law-abiding citizens who would not commit trespass in order to do their deplorable deeds. Rumour has it that there are apparently desperadoes who must have public footpaths in order to reach the site of their crimes.

Is there any truth in this? Must we all live in trepidation until winter rain and mud offer us some protection? These footpath hooligans must be worse than the normal sort, if what I hear is true. In all probability, panic will set in; the new bridges will be smashed, and barricades erected at every stile. The most innocent of walkers will be enough to send hearts aflutter, and in no time at all every walker of footpaths will become suspect. Eventually, perhaps, even highway-walkers will be regarded as potential highwaymen.

We must wait and see with what courage we can muster, but until I have proof that there is any truth in such dire forebodings I shall disregard them. What is more, when I have time, I shall walk the paths—if the prophets of doom have not succeeded in closing them.

# In Cobbett's footsteps

Whenever I travel the road from Gloucester to Huntley I think of William Cobbett, riding along on his horse on a September afternoon in 1826, feeling disgruntled because he had been unable to find a room for the night at Gloucester because of some 'Music Meeting.'

The 'Music Meeting' was the Three Choirs Festival, and Cobbett had no time for it; it was turning the Cathedral into an 'Opera-House,' as well as making travellers such as he bow very low and pay very high in order to find a room.

Cobbett was the last man to bow very low to anyone, so he rode on to Huntley where he found a room at (almost certainly) the Red Lion. He arrived there at five o'clock, probably stabling his horse, taking a meal and writing his diary before retiring to bed at seven-thirty with the 'intention of getting to Bollitree (six miles only) early enough in the morning to catch my sons in bed if they play the sluggard.' Next morning, he was delighted to find one of his sons already up and about and on his way to meet him; one of them managed a farm for a Mr Palmer at Bollitree, near Weston-under-Penyard, so Cobbett was often in the district.

A few days earlier he had been at Stroud, where he found hay fetching £6 per ton, twice the previous year's price. Later he visited Newent Fair, where the sheep trade was down a third on last year, Newent in those days being a centre for sheep.

Crops in Gloucestershire and Herefordshire were very poor that year; stock, wool, meat and cheese had fallen in price, but costs had risen, and dreadful ruin would 'fall upon the renting farmers, whether they rent the land or rent the money which enables them to call the land their own!'

On the Cotswolds he had seen the 'stout big-boned sheep,' and been told that many unsuccessful attempts had been made to cross them with the small-boned Leicester.

At Cirencester—which he described as 'a pretty large town, a pretty nice town,' he had once spoken in The Theatre. Most of the well-known actors and actresses of the time had also appeared there, and it had been patronised by the Duke of Wellington.

'Miserable country', Cobbett called the Ermine Street stretch from Cirencester towards Gloucester—until he came to 'Burlip Hill,' Birdlip, of course, and looked down into the Vale of Gloucester. 'All here is fine; fine farms, fine pastures, all inclosed fields; all divided by hedges; orchards a plenty,' he reported. 'Gloucester is a fine, clean, beautiful place.' The labourers looked well, and the girls working in the fields were not in rags, as in some other parts.

He did not think as well of Cheltenham: 'a nasty, ill-looking place, half clown and half cockney,' which appeared to be 'the residence of an assembly of tax-eaters.'

Cobbett had much abuse for 'tax-eaters,' for theorists who 'have a notion that there may be great public good through producing individual misery,' and for all those whom he described collectively as 'The Thing.' Yet William Hazlitt found him 'a very pleasant man—easy of access, affable, clear-headed, simple and mild in his manner.' Miss Mitford visited him when he farmed at Botley, and testified that domestically as well as publicly, he practised what he preached.

A farmer 'by taste and fact,' William Cobbett (1763–1835) was first and always a countryman. The son of a small farmer and innkeeper—'I do not remember a time when I did not earn my own living'—Cobbett believed in hard work and independence for himself and others.

A John Bull like figure, bluff, blunt, forthright, honest, his traits were to land him in prison. A lesser or more tactful man would have remained silent, but Cobbett was not a tactful man, and it was impossible for him to hold his tongue or stay his pen when he saw corruption, injustice, poverty or humbug.

He was opinionated, stubborn, awkward, sometimes wrong but more often boldly and gloriously right; a brave figure fighting the rising tide of industrialism and its bondage, waging a losing battle; a champion of freedom for England, especially rural England; and of the peasants, the poor and down-trodden and the dispossessed, of whom there were many at the time of enclosure. To him the question was: back to freedom or forward to slavery.

He was a champion of farmers, too, although he sometimes incurred their wrath; at one meeting many expressed what they would do to Cobbett if he were present, so he stood up to show them what manner of man he was. Good landlords earned his praise, others his abuse.

Farmer, gardener, nurseryman and seedsman, soldier, prisoner, exile, agitator, journalist, Member of Parliament (in his late years), conservative and radical, realist and idealist, a rebel and a traditionalist, Cobbett was big enough to encompass many roles. Above all, he was his own man, and prepared to fight for what he believed to be right.

With little formal education (which he scorned anyway) he became a master of clear, concise prose. He wrote many books, including *Rural Rides* (in print), *Cottage Economy* (recently reprinted), which is still the best self-sufficiency manual, and from 1802 until his death his *Political Register*, the most lively news-sheet of its time. His polemic style was devastating, as can be seen in his tirade against tea drinking: 'a destroyer of health, an enfeebler of the frame, an engender of effeminacy and laziness, debaucher of youth and a maker of misery for old age.'

He spent many years in America, where he also farmed, bringing Tom Paines's bones home with him; and he visited France and Ireland, as well as travelling extensively through England on horseback, with a shrewd eye for the state of humans, stock, fields and crops, and a keen one for wildlife and flowers; all of this is reflected in *Rural Rides*.

The noblest English example of the noble calling of the agitator, said Chesterton; but his agitation must not be confused with the role of the modern revolutionary. As Chesterton also said: 'After him radicalism is urbane—and Toryism suburban.'

H.J. Massingham called him the most English of Englishmen. More recently Professor W. J. Keith has written: 'History, when it has not been distorted in the interest of city-dwellers, has vindicated Cobbett. The importance of being on the winning side would have escaped him, and he would probably have condemned it as the product of an urban and false system of values. Once again, one feels, he would have been right.'

And when I travel the Huntley road, I salute you, William Cobbett.

# Gluttons for punishment

September 1986

What is worse than biting into an apple and finding a grub in it? The answer used to be finding half a grub, but now it could well be finding no grub at all. Each year a billion gallons of pesticides are sprayed on to British crops and soil, and at the last check a third of all fruit and vegetables were found to contain pesticide residues.

Beef cattle are injected with hormones which make them grow faster and produce leaner meat; pigs are fed antibiotics right through to slaughter; and the big poultry producers add growth-promoting antibiotics to the feed to the extent that an expert estimates that 80 per cent of supermarket chickens are contaminated by salmonella bacteria which *must* be killed by thorough cooking.

Pesticides collect in the outer layer of the wheat kernel, and Ministry scientists have admitted that the amount found in three-quarters of wholemeal bread subjected to tests have exceeded the maximum residue levels of the World Health Organisation.

We are told to eat fruit and vegetables, lean meat and poultry and wholemeal bread, and it seems sound advice; but we are rarely told that much of it may contain possibly dangerous toxic substances.

Perhaps even some nutritionists are unaware of the inherent dangers, but they no longer have the excuse of ignorance, for in *Gluttons for Punishment* (Penguin, £2.95) James Erlichman probes deep into the chemical warfare now being waged in agriculture and horticulture. It is a disturbing and startling book in which science and horror fiction become fact.

Mr Erlichman, chemicals correspondent of *The Guardian*, calls it a book about greed and addictions. A lot of people make money when our crops are soaked with excessive pesticides and our animals are dosed with too many antibiotics and hormones, and the chemical companies are having a field day. The big food chains and supermarkets deserve a share of the blame, too, for their demands for cosmetically perfect produce and lean meat at the lowest prices must force even reluctant farmers to reach for the medicine kit and pesticide spray.

The Government and its scientists insist that we are safe because all the drugs and pesticides approved for sale have been thoroughly tested, but James Erlichman argues that we are in fact the guinea-pig generation. He states his case clearly and concisely, and after reading his book I cannot see how anyone can feel safe unless his food is grown organically, or until there are far greater safeguards.

Sir Richard Body, the MP currently chairing the select committee on pesticide and health, is certainly not writing off its implications. 'This book is about life and death, and deserves to be read and considered very carefully,' he says.

It used to be said that an apple a day kept the doctor away; today I would far prefer to keep most apples away.

Pesticides, hormones and antibiotics make our food look better or grow faster and presumably they produce more of it too, and at lower costs. Yet we have embarrassing and expensive surplus of food, so who benefits? The farmer? The consumer? The food companies and supermarkets? The multinational drug and chemical companies? We all eat food, and after reading Erlichman's book I can only conclude that we are gluttons for punishment indeed.

# Silence, stubble and wold sheep

## September 1983

A hot afternoon in July, and the sun beat down relentlessly as I turned the last field of hay. When I had finished I stopped the tractor and sat in the shade of the hedge for ten minutes. Silence; all was still with no breath of wind, only the sun above and the heat shimmering, and cattle standing quietly in a group in the next field, lazily flicking their tails.

All was peaceful; nothing and nobody doing anything. But one only has to stop still and do nothing for a few minutes in the country to see and hear and learn a lot. Butterflies flitted busily to and fro, a kestrel hovered in the sky, and a cock and hen pheasant strutted importantly a few yards distant. Innumerable small winged creatures were intent on their business, beetles and a host of other creatures rushed, scrabbled and crawled industriously among the hedgerow grasses, regardless of the afternoon heat, and small birds fussed in the hedge.

Far from the sight and noise of other humans we think the countryside deserted, quiet and placid, but to the countless number of creatures of field and woodland it is as crowded and busy as towns are to us—and even more fraught with danger. Fox prowling after bird, bird after worm, kestrel or owl pouncing on mouse, stoat pouncing on rabbit and so on. The bird song we find so delightful, too, may not be so to other birds.

The wide open spaces to us are a congested area to them, each tract of territory to be defended against invasion; species fighting species, species preying upon species, homes to build or dig, young to be reared and taught to fend for themselves. Then there is the danger from man, who may quickly destroy their townships with sprays, chainsaw or bulldozer—or who can exterminate home and life for myriads with a single match in a field of stubble.

When haymaking was finished we went to Bibury, seen by William Morris as the most beautiful village in England. We visited the museum in Arlington Mill, admired the fabled Arlington Row, and then Harley and I sat by the river to watch the ducks and trout. Harley is my grandson, who will be three this month. Weed like green hair flowed, and brown weed like fox brushes waved and bobbed in the clear water. Big ducks and little ducks swam and scurried about the water, and big trout and little trout gleamed and darted beneath.

'There's a big one,' I said, pointing, but the fish had gone before I had finished speaking. 'Yes,' said Harley solemnly, but whether he had seen it I knew not; Harley, I suspect, often says yes to humour me. With the wisdom possessed only by the very young, he knows that grandfathers need to be humoured from time to time. People talk of the generation gap as if it is something to be deplored, but as far as Harley and I are concerned it binds us in understanding and firm friendship.

Other ducks just sat placidly and idly on the water, other trout just stayed beneath, but in that clear water I saw that even to stay put, seemingly lazy, both duck and trout had to be busy. The webbed feet of duck, the fins and tail of fish are continually active. Oh, idle but restless life of duck and trout, of weed in water, of reed shaken by gentle breezes ...

On the way home harvest was in full swing on Cotswold's acres, the sight of each combine bringing a delighted cry from Harley, followed by an imitation of its sound. For a moment, he, too, was a combine; such is the versatility and imagination of a small boy to whom the whole world is a wonder of delight.

And to my delight, no sign of straw-burning; instead, field after field of large round straw bales looking like giant toilet rolls—and more sheep than I have ever seen on Cotswold. Perhaps it is becoming shepherd's country once again.

# Buying a couple of pigs

September 1977

The notice on the tumbledown wall read: 'Pure New Honey'. The gate was off its hinges, and I had difficulty in opening it. Grass grew in the cracks between the paving stones that led to the front door, and the stones were green with slime.

The drab paint on the door was blistered and peeling. I knocked. At a window a grimy curtain twitched, and after some scuffling the door opened wide enough for a head to poke out.

'You'll have come about the pigs,' said the head. I nodded, and the door opened wide enough to reveal an old woman. Grey, that's the only way to describe her, not only her hair and clothes, but her whole appearance and demeanour—just grey. I followed her through the neglected garden, rotting cabbages, sprout stalks, brambles, docks and nettles. The plot was littered with upturned rusty buckets, presumably for forcing rhubarb. We clambered over an old iron hurdle, and I wondered how the old woman managed. She must have noticed my look, for she spoke again.

'Yes, twice a day I've got to carry buckets of pig food over here. It's not good enough. He promises to put me a gate, but as you see, he doesn't.'

I agreed that it did not seem good enough, and a gate was certainly needed.

We were now out in a field littered with all manner of items, mostly junk. She told me that 'he' was not at home, but was most likely in the Green Man or the Wagon and Horses. In the middle of the field was a shed, beyond which the field sloped steeply up to a high hedge. It surprised me slightly when a bicycle suddenly came hurtling over the top of this; even more so when a figure appeared crawling through the undergrowth.

It stood up to be revealed as a small man with very baggy trousers, and a large moustache, who promptly started to run down the slope, his knees coming up almost to his chin.

'That's him and he's drunk,' I was told.

With deepest blue eyes twinkling, he passed us smiling and waving. 'Can't stop, see you soon,' he sang out, disappearing behind some brambles.

In silence we awaited his return, but instead he ignored us and went straight inside the hovel. After a few minutes I heard his voice, and thinking it was an invitation to go inside, I did so.

It was gloomy in there, and I couldn't see him at first, but as I grew accustomed to the darkness I spied him lying alongside a big saddleback sow, playing with her and talking to her piglets.

'There then, my little beauties. Oh, my little darlings.' He took not the slightest notice of me, but continued to fondle the pigs.

After a while, I grew rather tired of this for a caper, and told him I had come to buy a couple of the young gilts. He looked up at me with a radiant smile: 'Ah, but I don't know as I want to sell 'em.'

'But you advertised them in the paper,' I exclaimed.

This had him nonplussed for a while, but eventually he spoke again.

'There's that old sawmill of mine lying idle. Now, you do know, that oughta be working.'

This new tactic, in turn, set me back a little, but, as briskly as I could, I returned to the business in hand, and, after more inconsequential mutterings, he spoke of the pigs again.

'She's got ten pigs: I think she had eleven and lay on one, because I smelt something nasty under the straw, but I never looked. There's four gilts, and I've sold two of 'em to the Green Man. You can have the other two in a fortnight for eight pound apiece.'

With a dreamy grin, he settled back in the straw, lost interest in me, and I left him uttering endearments to his family. It was good to be back in the air again.

On my way back I passed his wife feeding her fowls. She asked if I'd bought any of the pigs.

'Two of the gilts. The others are sold to the Green Man.'

'Oh dear, I wish my pigs weren't going there. They aren't proper sort of people. I might have guessed it; but they're not what I call proper people at all.'

# Butterflies

September 1978

Many of our butterflies are in decline, and in danger of extinction. There are several reasons for this which are beyond our control; butterflies have their natural enemies and diseases, and adverse weather conditions, and the Large Blue, which once bred on the Cotswolds, was dependent upon a particular ant in its life cycle.

One of the principal causes for the dearth is undoubtedly modern agricultural and forestry practice; the destruction of hedgerows and woodland, the widespread use of herbicides and pesticides, the cutting of road verges, and the over-assiduous efforts of the best kept village competitions.

It is odd, in an age of man-made litter and pollution, that so many wild flowers, grasses and weeds—and nettles, thistles and brambles are excellent butterfly plants—should be eliminated on the grounds of tidiness.

There are signs of a more enlightened attitude to wildlife, however. The high cost of chemical sprays may lead to their more discriminate use. The need to economise has caused local authorities to leave road verges untrimmed. Wild violets, primroses, tall grasses, Jack-by-the-hedge, hops, knapweed, moon daisies, yarrow, vetches, and white clover are some of the wayside plants beloved by butterflies.

Gardeners, too, can grow shrubs and plants to attract them. Everyone knows about buddleia—the butterfly bush—though I find the late-flowering pink sedum attracts far more butterflies. It is easy to grow, and slips taken from the parent plant soon make fresh clumps.

But it is no good attracting butterflies without taking account of breeding and hibernation, and the following should provide for both, so long as there are also tall grasses and a patch of stinging nettles close by: Aubretia, arabis, buddleia, candytuft, catmint, fennel, golden rod, holly, ivy, lavender, lilac, moon daisies, phlox, sedum, sweet rocket, sweet william, virginia stock, and wild thyme. An old shed makes an ideal winter quarters.

These, and the wild plants mentioned earlier, should bring butterflies and delight to any garden. The Cabbage White is the only species that causes any damage, and such is the dearth of the benevolent types that several firms supply butterflies: Thompson and Morgan, the seedsmen of Ipswich, can be contacted for pupae and further information.

# Of men and mangolds

## September 1979

The winter was loth to leave us, and spring and the cuckoo were late in arriving. Cattle fodder dwindled, the grass was reluctant to grow, and the prospect of a good hay crop appeared unlikely.

Rough, cold winds shook the buds of May and made us shiver; torrential rain at the end of the month flooded the fields and roads; even at the end of June the hay would not make. We had had no summer by then, and there seemed little likelihood that we would.

Then, in July, the elements made amends. The sun shone boldly day after day; the grass had grown, still young and leafy, and a few days after cutting it was transformed into blue-green hay and baled. It was a pleasure to load and stack it, half the satisfaction being the thought of the cattle's pleasure in feeding from it next winter.

And, of course, the better the hay the less it costs to make. It is the rain-soaked crop of little value that costs the most, with that interminable turning, work and worry. This year we made the best of hay in the best of haymaking times.

The dry weather continued, the pastures grew brown. We needed rain, there was no doubt of that; and yet, and yet … Cattle always do better in a dry summer, even if the pastures are bare, so long as they have shade and water. Drought has never meant dearth in England yet—although it came close to it three years ago.

This year I had my first taste of modern, fully-mechanised silage-making. I had made silage before—when few other farmers did—for seventeen years, but there were always problems and I stopped. Within a few years almost every other farmer was in on the act, with improved harvesters and techniques. I agreed that hay and concentrates added up to a dull diet for cows in the winter, for they need something succulent. But against the trend, I thought I would return to mangolds.

A big dung heap was piled in the field in readiness for the crop, but before it could be spread and the field ploughed, the bad weather came. That is partly why we did not grow mangolds—but the real reason was because my courage failed me.

I quailed at the thought of hauling several hundred tons of them on our heavy land in a wet autumn. I saw tractor and trailer bogged down in mud, wheels spinning helplessly, mud flying, trailer axle-deep. If I had had a horse and cart it would have been different.

But in our few hectic days of silage-making, how I wished I had taken the plunge. The process was quick, efficient and lacked the hard physical labour that mangolds would have demanded—but it was all done from the seats of tractors, and everyone was an island, lapped by the roar of machinery. There was no let-up; no time, less chance for a talk. No companionship. With mangolds there would have been both of these sweeteners.

I still wonder whether the days of these modern, powerful machines are numbered; I wonder, too, about efficiency—a much-used word, and an ill-used word, which all too often implies something unpleasant in the offing. Until happiness finds a place in its calculations it cannot be a very satisfactory yardstick for measuring the workings of people.

# Diary Dates, dire dates ...

'September blow soft till fruit be in loft;' but many apples, and most pears, are not ripe for picking until well into October. This, like much else in Thomas Tusser's farming calendar of 1577, is enough to prompt the thought that the seasons have grown later—until it is recalled that Tusser was writing before our calendar was changed. Four hundred years ago exactly, Pope Gregory XIII ruled that October 5 should be called October 15 to correct an error in the Julian Calendar—a ruling England did not accept until 230 years ago this month, when our September 5 became September 14.

This Gregorian ruling may have been a correction, but even today it still upsets many of our festivals and customs. Take May Day, for example. In the Julian Calendar it fell on the day we now call May 13, a throwback to the medieval day of May poles, garlands of hawthorn blossom and fertility rites deriving from Celtic times, when the year was divided into summer and winter only.

The first day of May was the first day of summer, the day stock was turned on to the commons. The May pole, far removed from the Victorian revival, and the gathering of hawthorn, 'bringing in the May,' were part of the fertility rites. People then lived closer to the earth, were in every sense more earthy, and the fertility of animals and stock was of concern to all. Now the hawthorn, oblivious to Pope Gregory's ruling, still blooms on the May Day of old.

For the Celts winter started on October 31, the day we call All Hallows Eve, but because of the Gregorian correction we are several days too early. This was the day when stock was brought in, and bonfires which now burn on November 5 were lit to the dying sun.

The mid-winter festival we now know as Christmas Day is of Pagan origin, to celebrate the birth of the sun. Again we are now several days too early, yet we still expect the kind of weather more usual before the change of calendar. People born as late as the 1880s spoke of old Christmas Day as though that was the day upon which they once celebrated the festival. And old Christmas Day, with its subsequent period of festivities, would see us nicely through the real turn of the year and to the 'birth of the sun,' as it did in medieval times.

Talking of dates and anniversaries reminds me that it was just 250 years ago that there was that unpleasant business at Mitcheldean—a black festival, indeed.

On Saturday, April 29, 1732, Thomas Twibervile, a carpenter of Mitcheldean, did not open his shop. A childless widower, he lived alone and it was not until the evening of that day that alarmed townspeople broke into his shop and discovered him dead. His headless body lay on the floor of the workshop among the blood-stained shavings and sawdust, his gore-covered axe beside him. His severed head lay on a workbench with the brains dashed out and the skull smashed to fragments.

Eli Hatton, a weaver of Ruardean, a reputed deer-stealer and Sabbath-breaker, was arrested on suspicion. He was kept in custody until the inquest two days later, and was then taken to the prison of St Briavel's Castle to await trial at Gloucester Assizes. Several hundred guineas, according to Twibervile's account book, were missing, and extensive searches failed to find the money, then or later.

At the trial, Eli Hatton was found guilty and condemned to be hanged on Pingry Tump, Meane Hill, near Mitcheldean, and his body to hang in chains. Protesting his innocence, he proclaimed that it was all a plot against him, but he did admit that he was a Sabbath-breaker, and that it was usual for young people of the district to work all week and steal deer on Sundays.

On September 4, 1732, at 7 a.m., he was taken to Meane Hill, where ten thousand people gathered. At that time the combined population of Mitcheldean, Littledean and Ruardean amounted to only 1,550, and the rest of the Forest of Dean was sparsely populated. Even if the number of spectators was greatly exaggerated, there must have still been a great journeying from a large area.

On the crest of the hill at Pingry Tump stood the gibbet. Eli Hatton was forced up the ladder and his hands tied. The spectators removed their hats, the noose was tightened, and soon Hatton was swinging lifelessly. The body hung in chains for weeks, and no one went near Pingry Tump.

Then came the rumour that the flies which fed and bred on his remains were coming down to Mitcheldean and tainting the meat on the stalls of the market. In an effort to save their trade, the butchers went one night and removed Eli and buried him in a secret grave. No one else even dared venture near Pingry Tump for many years, though, especially at night; and the stump of the gibbet came to be known as Eli's Post.

# Nine gallons a day—hats off to them

September 1980

Two large prints, almost one-hundred-and-eighty years old, hang on a wall in our dining room. One is a picture of a Suffolk bull, the other a Suffolk cow.

The Suffolk, or Suffolk Dun, is now extinct—one of the last herds was at Cirencester—but it is one of the ancestors of the Red Poll. Arthur Young (1741–1820), the first secretary to the Board of Agriculture, called it the little mongrel breed, and spoke highly of its milking abilities.

Most dairies of any consideration, he said, boasted cows that gave nine gallons of milk daily at the height of the season in early June and six gallons for most of the season. Whole herds would average a daily yield of five gallons per cow for three or four months.

Other writers gave similar instances of high yields of good-quality milk from Suffolk cows; and these were obtained without the aid of concentrates and other forms of scientific feeding or management.

Arthur Young, William Cobbett and other writers of the period also noted wheat yields of two tons per acre. They must have been exceptional, but nevertheless, they, too, were gained without the help of artificial fertilisers, chemical sprays or any of the other aids we have today. With some of today's highest-yielding wheat varieties producing over three tons per acre, we have reached the absurd position where it takes more units of energy to grow the crop than it eventually yields.

Powerful and heavy machinery can also be self-defeating, its use leading to the need for even more powerful and heavier machinery. I doubt if some land once ploughed by a pair of horses could still be cultivated in the same way today—or even by the small, light tractors popular thirty years ago.

Gigantic machines have compacted the soil; and a reliance on artificials and a lack of humus have made it more difficult to work, especially on large farms which grow crops of cereals year-in, year-out, and where the straw is burned instead of being returned to the land as dung.

Talking of which, a conservation adviser says that straw-burning is a necessary part of harvest; it will destroy disease that chemicals cannot destroy. Perhaps it does, and under monoculture it may be necessary—but monoculture itself is alien to nature and to good farming.

This is but one example, but it seems that some conservationists are in danger of forgetting the meaning of conservation, and that a return to husbandry would do far more for the cause than their support for some of the doubtful practices of modern farming.

When husbandry was the rule, the countryside was enhanced and its wildlife secure. Husbandry and conservation go hand in hand; agri-business and conservation do not. But it is possible to grow food without destroying the countryside or its wildlife, without bullying the land, and without animals being unduly exploited.

The latter point is important, too. When the housewife buys eggs labelled Farm Fresh, they may be fresh and from a farm; but does she realise they may have come from a factory on that farm where the hens are confined two or three to a cage?

# Farm sales

Sun or rain, farm sales invariably attract a large crowd; some come to buy, but many more just come to stand and stare, to listen and talk. Farm sales are usually late in starting, often half an hour after the advertised time, but the problem of parking, the delays and the waiting, the 'hanging about'; none of these deter the salegoer.

Some are habitues, attending every sale for miles around; some seem to exist only for sales, living in a kind of limbo between times. But on the whole they make a good-humoured gathering, the many who come to view and the few who come to buy, all ready to laugh at the auctioneer's slightest quip.

Apprehensive livestock penned and labelled; machines, implements, equipment and tools also labelled and arrayed in the home meadow, and collectively described as deadstock. This ranges from the nearly new to the worn, the damaged, and the broken, and finally degenerates into heaps of useless junk. Surprisingly, even the junk finds buyers, and often at tidy prices.

Seeing the deadstock awaiting the auctioneer makes one wonder what one's own collection would look like thus displayed—an eventuality that can be prevented only by the happy combination of good fortune, a son, and security of tenure. A sobering thought, when you realise that it's a collection of a lifetime lined up there in the meadow; several lifetimes if the farm has been in the occupation of the same family for generations.

But despite the auctioneer's brisk manner and jokes, and the good-natured crowd, I find farm sales melancholy occasions, invariably with a background of failure, ill-health, old age or death. Saddest of all, when a herd is sold. A herd of cows, as distinct from a collection, is the result of a lifetime of care and skill, of heartbreak and triumph. Indeed, one man's lifetime is all too short to create the perfect herd. And then, suddenly, with the rap of the auctioneer's hammer, it's all undone, dispersed, animal by animal.

Late that same day, when auctioneer, farmer, dealer and onlooker have departed, when the last animal has been driven into the cattle truck, and when all but the useless junk has been cleared, a silence descends on the farmstead. My only hope then is that, soon, fresh animals will rustle the straw or graze in the pastures as before.

# Kale as blue as the sky …

## September 1984

By the time this appears it may be raining cats and dogs. I must look in *Brewer's Dictionary of Phrase and Fable* to see why 'cats and dogs'—but not now. Brewer's is seductive, more seductive than most dictionaries; look up one word and you go on looking and looking, and sometimes you even forget what caused you to open this beguiling volume in the first place. Now I must concentrate on this piece for September, and not go browsing through Brewer.

It is the last Sunday in July, and we have had no rain since the end of May. Some showers then were just enough to start the kale growing—and the weeds. We sprayed and killed the weeds, leaving a nice plant of kale, but with the prolonged drought it has not been able to grow and is now as blue as the sky which pours down its relentless heat day after day.

The mangolds, on the other hand, have hardly been affected by the drought—and the weeds among them have been affected by neither drought nor spray. I only wish we had the time to hoe as we once did. I wish we had a horse and a horse-hoe too, and by the time we come to harvest the mangolds it will probably be wet, and I shall be wishing we had a horse and cart.

All our hay was made and gathered early, without a spot of rain falling on it; I do not expect to see many more seasons when that can be said. It is supposed that there is a break between hay and harvest; often enough there is not, because rain has delayed haymaking and there was not this year, either, with all the dry weather. The winter barley was early—in ear on May 5, almost a record, surely—and it ripened quickly because of the drought and heat, neither of which did it any harm. They are saying the yields of wheat will be lower because of the drought, but I find it hard to believe.

The berries of corn are small, but most modern varieties seem that way, and the seed-corn we get is certainly so, far too small, in my opinion. Most seed merchants disagree; small seed will produce plump corn given the right conditions they say. Perhaps—but if so, why wasn't the seed-corn they sell me given the right conditions; and if it was, why is it so small?

Their answer is that it does not matter, and that the seed has been tested for vigour and reliability. So it may have been, but years ago it would have been scorned as tail corn. There was a time when farmers saw samples of seed before buying; reasonably and sensibly, they took the view that as they were buying and planting the seed, they would be the judges. They would not have been impressed by a lot of jargon and gobbledegook, and would not have been fobbed off with anything the merchants cared to foist on them.

Yesterday we were hauling wheat straw we had bought from a friend, and he told me that this autumn he intends to insist on seeing samples before he buys his seed-corn. In other words, he is going to judge and choose his seed himself—which is exactly what the merchants do when they buy his crop for milling or feeding.

The long drought has opened up great cracks in the ground, which has made on-the-field repairs to machinery more frustrating than usual. All too easily nuts and bolts disappear down the cracks, replacements have to be found—and that means back to the buildings for more. Even then it is not always easy to locate just what is needed. If only nuts, bolts, springs, washers and split pins could be produced like corn or roots, I would grow half an acre.

At the moment there is no sign of rain; we have had signs of it earlier, and there have even been reports of the real thing in other places, but it has passed us by. I never listen to or read weather forecasts, so I do not know what the met. men have to say. All I know is that the pastures are brown and crisp or brown and bare, baked hard as iron and cracked—but the cattle look well and seem contented in spite of the absence of grass; they always do, so long as they have plenty of shade and water.

Milk yields, of course, have shrunk, and are shrinking daily. It does not matter, since the milk quotas have seen to that, but the quotas themselves matter. Hastily and unjustly applied and ill-conceived, they are a monumental blunder that will mean the end of many small milk-producers, whatever the weather.

# Barton Fair

September 1985

Barton Fair, which used to be held at Gloucester on September 28 and is now on the last Thursday of the month, is one of the few surviving market fairs in Gloucestershire, Stow being another. There are, of course, plenty of pleasure fairs, which are often our sole reminder of what were once important days in the county's rural life and economy.

Thomas Ridge, in his *Agriculture of Gloucester* of 1813, lists twenty-five market towns which had one or more fairs every year; most of them sold horses, cattle and sheep, while some also handled cheese, perhaps, and others wool. Campden and Stow sold linen and cloth.

Charters to hold fairs were granted by the Crown, and while many charter fairs had their origins in Norman times, others pre-dated the granting of charters. The right to hold a fair was a great privilege—and a source of income to its holders, since it allowed tolls to be charged on all livestock and goods sold.

The charter to hold Barton Fair was granted to the Abbey of St Peter, Gloucester, in 1227 and was later passed to the Mayor and Burgesses of the Manor of Barton Abbotts. Horses, cattle, sheep and cheese were sold in certain streets of Gloucester until 1821, when a local Act authorised the City Corporation to set up a proper place to hold markets and fairs. A new market was built in 1855 and enlarged nineteen years later; covering about four acres and today known as the old market, it stood on what is now the bus station. The present market, off St Oswald's Road, on the fringes of the city centre, was built in 1955.

This month's Barton Fair at Michaelmas, the start of the farming year, was also a hiring or mop fair. Farm and domestic workers seeking employment would attend, bearing the insignia of their trade. Shepherds wore a tuft of wool, carters a piece of whipcord, cowmen some hair from a cow's tail, and domestic servants carried a mop. People engaged would be given earnest or hiring money—and unless this token was returned, the contract, for good or ill, was binding on both parties for a full year. In some places there were later fairs, known as runaway mops, where those who had not found jobs on the earlier day were given another chance.

The rest of the fair day was given over to entertainment, the only holiday many workers had, and that is how the pleasure fairs developed. I imagine the practice of hiring at fairs must have been obsolete by the beginning of the century.

With the end of so many local fairs, Barton must have gained greater importance. Gloucestershire is divided geographically and thus agriculturally into three parts, Cotswold, Vale and Dean, although in this context the last would more properly be called West of Severn. Barton Fair was a focal point where, once a year, all three parts of the county converged and met and became whole, the farmers of Gloucestershire. They would buy or sell, talk and make merry—and in those days farmers, dressed in breeches and leggings, could not possibly be mistaken for anything else.

Gloucester would be invaded by them, men from the hills and the vale. The market would be choked by horses and sheep—especially sheep—redolent of them, and loud with the sound of them, the noise of hand-bells, of talk and of auctioneers. Resplendent in grey bowler and yellow waistcoat would be Critch Pope, chanting: 'How much am I bid? How much am I bid?'

Old men who rarely stirred from their farms for most of the year would do their best to be at Barton Fair. At almost ninety years of age my grandfather still went. With eyes atwinkle in the market place he would say with pride: 'I haven't missed a Barton Fair for over seventy years. There's not many here today who can say that.'

After inspecting the sheep—his lifelong interest and joy—and having met an old friend he had not seen since the last fair, he would retire to one of the many pubs which surrounded the market; all of them, of course, would be doing a roaring trade. It was a great occasion for my grandfather and men like him, a day for meeting old friends, for drinking and talking of old times. Yet it must have had its melancholy side, too, for each year another old friend would be absent. At such gatherings, at such an age, the ghosts must have outnumbered the living.

My grandfather has been dead for almost forty years. Critch Pope is gone, as well, and so are most of the farmers who wore breeches and leggings, and to whom farming was a way of life as well as a business. The old market in the town is no more, and tides of sheep and farmers no longer flood into Gloucester on Fair Day. Absent, too, is much of the romance, the bustle and chaos and the colourful characters, just as much a part of the past as the cheese and the hirings of the fairs before them. Yet large numbers of sheep are sold on the last Thursday of September at the new market on the outskirts of the city; and it is still Barton Fair Day to the farmers of the hills and vales of Gloucestershire.

# The Jovial Collier

## October 1985

The *Jovial Foresters*, a song of the Forest of Dean, was a huge hit in London during the 1860s; sung by Jolly John Nash, dressed as a collier complete with 'tommy bag' and water can, it was a great success in the music halls.

Jolly John Nash came from Coleford. He played at the Theatre Royal, Gloucester in 1866, three years before Henry Irving's first visit as a 'secondary actor,' but I must confess I know little else about him. It was no doubt he who altered the last word of line seven, verse two, to 'Queen'; in order to make it appropriate to Victoria, but as 'King' was the obvious rhyming word, it is clear that the verses were written during or before the reign of William IV.

Writing in 1926 Mr A.W. Pope, a bootmaker and musician of Cinderford, recalled seeing a copy of the song some thirty years previously. It did not carry the composer's name, but he later discovered it was written by a William Trotter of Coleford, a fact subsequently confirmed by a nephew of Trotter; he had seen a copy in 1870, with an illustrated cover which apparently showed the engine house at Lightmoor Colliery, its window open to reveal a jolly crew and a table with jugs and glasses. Lightmoor Colliery dates from 1830, and its engine house is the only one now left standing in the forest.

The song was dedicated to Edward Protheroe Esq. He opened several collieries in 1827, and by 1831 he owned thirty-two Forest pits, including Crump Meadow Colliery. He also laid an extensive tramway system in the Forest, and built a substantial school on Cinderford Tump 'where the holly grew.'

The *Jovial Foresters* tune, again according to A.W. Pope, was later adapted and used as a brass band quickstep. The melody had probably descended from the old Scottish *Ballad of the Jovial Beggar*—'There was a jolly beggar an' a begging he was born …'—a song which has been ascribed to James V of Scotland. A later and undated copy published by the *Three Forest Newspapers* has an arrangement with an accompaniment for piano by Joshua Hatton.

On at least two occasions the song has been adapted for use as an election song in the Liberal interest. In 1868, when the Forest was in the West Gloucestershire Division, as it is again now, it returned two members. Colonel Kingscote and Mr S. Marling were the Liberal candidates, which led to:

For we are the Jovial Foresters,
We'll fight with heart and soul,
Kingscote and Marling hand in hand,
Shall head the Forest poll.

Later, in 1892, when Sir Charles Dilke first stood as Liberal candidate for the Forest, they sang:

We are the Jovial Foresters,
Sir Charles shall be our man;
We'll send him back to Parliament
To help the Grand Old Man.

That election song, which had thirty-four lines, was sung at all the Liberal meetings and at some Conservative ones, too.

The Grand Old Man, William Ewart Gladstone, may not have viewed the return of Dilke so enthusiastically as the Foresters, but the last line of the full version, 'And shall long our Member be,' was certainly accurate; Sir Charles remained MP for the Forest until his death in 1911.

# The Jovial Foresters

For we are the Jovial Foresters,
  Our trade is getting coal,
You never knew a Forester
  But was a hearty soul.
Tho' black we are when at our work,
  You'd take us for some smoking Turk;
When that is done we're ripe for fun,
  To laugh and chat with anyone.

*Chorus*:
For we are the Jovial Foresters,
Our trade is getting coal;
You never knew a Forester
But was a hearty soul.

Among mankind there miners are
  Of every degree,
But he who undermines his friend
  Is far more black than we.
He's black at heart without a doubt,
  And there's a stain will ne'er wash out,
But we are true to Church and King,
  Therefore let us merrily sing.

For we are the Jovial Foresters ...

As to the Church, it must be owned
　　To that we are no foes,
For while we undermine the Dean
　　We warm the Bishop's nose.
In vain may Chloe turn the spit,
　　Nor could the cook your fancy hit,
Nor for the Mayor the feats prepare,
　　Were it not for the toil and care

Of us, the Jovial Foresters ...


The courtier undermines the State,
　　As to the doctor, he
Your constitution undermines
　　But to prolong his fee,
The lawyer undermines your purse,
　　But none of them can work like us,
For we are bent with full intent,
　　To raise the coal and give content;

For we are the Jovial Foresters ...


To raise the coal is all our aim;
　　For that the soldier fights;
For that the player acts his part;
　　For that the poet writes.
For that the barrister doth plead;
　　For that the ladies give their aid.
The parson prays the coal to raise,
　　We're colliers all in different ways;

But we are the Jovial Foresters ...

The Foresters then drink a health,
　　Wherever they may be,
For they are the lads who free from care
　　Through life jog merrily.

May plenty reign throughout the land,
    That bread they may have at their command;
And this I think, while you've the chink,
    You never will refuse to drink

To us the Jovial Foresters ...

# The obstreperous October onion …

## October 1979

It is a still morning in October with a nip in the air, and I'm harrowing the ploughed earth in preparation for winter corn. It's a time of hope, this season of planting. It's like writing in a way, every time the pen is picked up one hopes to produce something good. Without this hope a crop would never be sown, a line never written.

There's another parallel between farming and writing; you plough the same soil, use the same words that have been used for generations; which reminds me that there's next month's article to be written. I haven't a thought about that yet, and as I harrow I must put my mind to it.

The earth is clothed with a fine net of webs, the work of a myriad, which the harrows are destroying. I think of Burns's lines: 'I'm truly sorry man's dominion should destroy nature's social union.'

The union of white clover and ryegrass; of leaf and flower, of bird and beast. We know so little and destroy so much. But nature too can be red in tooth and claw. Yes, but I should be thinking about next month's column …

Soon the leaves will be turning, and all too soon golden October, dear October will be gone, although it's only just arrived. November will be here; dripping, damp November, foggy November—and its mud—although in its defence it does not always live up to its reputation. Thomas Hood has a lot to answer for with his:

No fruits, no flowers, no leaves, no birds—November.

But for me on the tractor it is still October—with that article to be written tonight. Ah but November's dear to the farmer, too, dark evenings by the fire, no nagging thoughts that you should be working; unless a cow should be calving, perhaps, a life of ease after tea.

Up and down the field with the harrows, up again. I think and think, but still no thoughts for November's column.

Question: can one think about a thought? If ideas could be plucked from the air like apples from a tree, how easy writing would be; almost half as easy as some already think it is. But ideas, if they come at all, come from some mysterious nowhere. They may come as gently as floating thistledown, suddenly as a hawk to its prey, quietly as a mouse—or they may not come at all. They may never be coaxed, demanded never. One could try, I suppose, as Glendower boasted he could call spirits from the vasty deep, but as Hotspur replied: Will they come?

My neighbour is already planting, I see, and his cows have started grazing kale. How long will his kale last? Or my mixture of rape, kale and turnips, I wonder, or our other stocks of winter fodder? Will Daisy calve tonight?

Half the field is harrowed, I've thought and thought, but nothing has come. You'd think, wouldn't you, that sitting all alone on a tractor simply harrowing a twelve-acre field would be conducive to thought, ideal for ideas? Well, you'd be wrong; on this particular day, at least, I can assure you it's definitely not.

My neighbour's stopped in mid-field; something must have gone wrong. It's his cows, they've broken through the electric fence ... and just as I'm least expecting it, something comes into my head; it's not a thought, less still an idea. It's a simple jingle:

In every good cook's opinion,
No savoury dish without an onion.

I finish the field, and it's still in my head. Even if anything else did want to come, it couldn't. My brain is still full of:

In every good cook's opinion ...

It's afternoon, milking time, teatime; writing time:

In every good cook's ...

It's no good, I can think of nothing else ...

# Dark clouds of ignorance

October 1983

By the beginning of August we saw great clouds of smoke on the Cotswold hills, some twenty or more miles away. Soon there were blazing fields in the Vale; not far from here I saw a score of scorched fields, with burned hedges, trees and bushes.

Straw-burning on the Cotswolds has become an established practice—in spite of my premature optimism of last month—and now, with larger farms and the increase in specialist corn-growing as a result of current agricultural policies, it seems that it will become standard, too, in the traditional livestock district of the Vale.

Agri-business says there is a surplus of straw, without adding that there is also a surplus of corn, for which the public has to pay in one way or another, whether it likes it or not. White straw crop follows white straw crop, and the corn is sold for a guaranteed price; the straw is sold, too—or burned because there is no livestock to convert it into fertility.

There was a time when such a system would have been condemned, but today science has provided techniques to enable farmers to ignore the old code of good husbandry. The bag has replaced the dung heap, and sprays and straw-burning help keep at bay the disease one would expect to result from monoculture. So far science has been able to stay a jump ahead of nature—but nature has a habit of hitting back if it is flouted too much and too often.

Science has done wonderful things for farming. It has brought better and heavier crops, and taken away much of the drudgery, as well as giving the farmer a great deal of knowledge in place of superstition. With that knowledge, though, it has bred a great deal of ignorance and arrogance, and it may well be that science has yet to learn, and then to teach modern farmers, the lesson that superstition already knew. The old-time farmer knew by instinct that you could not take more than you put into the land. Many modern farmers think—they can—and do—take more than they give. So far they have done so fairly successfully, and only time will tell if they are right.

About twenty years ago that eminent agricultural scientist, Sir George Stapledon, declared that he now saw what he saw because he no longer studied science or adopted the techniques of science. He had come to realise that facts and factors as such meant precisely nothing in agriculture: 'It is their mass inter-relationships and inter-actions that mean everything. Man, in putting all his money on narrow specialisation and on the newly-dawned age of technology, has backed a wild horse which, given its head, is bound to get out of control.'

The disgusting and widespread practice of straw-burning is just one of the results of narrow specialisation, and technology has divorced man and animal from the land in a way that has led to this wasteful practice. Straw-burning is one of the more glaring symptoms of unbalanced farming; not the occasional 'burn', which may be justified, but regular burning.

There was a time when farming enhanced the countryside and did not endanger its wildlife—but who except a 'modern progressive farmer' would say that is still the case? Of course a lot of farmers still do care about more than how much they can drag out of the land, but today's policies and economics are ensuring that they are diminishing.

Unlike the large specialist arable farmer, the small mixed farmer finds his way of life increasingly difficult as his income grows smaller. If, or more likely, when, his small farm is sold, it is usually absorbed into a bigger one, with the farmhouse sold off and the hedges grubbed. Mind you, the large-scale farmer is not always deaf to the voices of conservation. He sometimes plants trees—grant-aided, of course—and conservation bodies are becoming adept at arguing their case in terms of big-business cost-effectiveness.

It is not the new knowledge that frightens me; it is the new ignorance, and the arrogance bred of it. As Sir Peter Scott said: 'Be concerned about all life on earth; about the beauty and diversity of animals and plants—not forgetting that it's their world, too.'

If we all remember that, there is a fair chance that we shall still have a world worth living in—and that our great-grandchildren may never need ask: 'What *was* England's green and pleasant land?'

# The sour taste of sugar beet

## October 1982

This month should see the start of the sugar beet harvest. Sugar beet, which grows wild in the south of France, was developed as a cultivated crop by the French during the Napoleonic wars, when the British blockade prevented them from importing sugar from the West Indies.

It was introduced into this country in 1909, and the first English sugar beet factory was established at Cantley, Norfolk, seventy years ago. Now it is grown extensively in the eastern counties and elsewhere, including our neighbouring Herefordshire. I do not think it has ever been a really popular crop in Gloucestershire, but during the last war, when our supplies of sugar from the West Indies were erratic, many farmers round here were ordered by the War Agricultural Committee to grow it.

Perhaps my dislike of sugar stems as much from those days as from personal taste. All those weeks of hoeing; singling with hand hoes, horse-hoeing, more hand-hoeing, more horse-hoeing. The singling was tedious; beet seed, like mangold, grows in clusters, but unlike mangolds, beet seedlings twist and cling around each other, making singling difficult and tedious.

Day after day, week after week, we spent back-breaking hours hoeing the beet. One day a smallholder not directed to grow the crop leaned on the gate and watched us. 'Will it ever pay?' he asked. Without straightening his back my father replied: 'It's damn well got to pay,' and continued with his hoeing.

In the autumn a bouting plough was used to lift the beet, but this operation was more of a hope than a realisation. The beet, with its long tap roots, clung tenaciously to the sub-soil. The beet held most of the advantages, while man just held its leaves and pulled.

Eventually man won, and laid the beet in rows across the field, after which the leaves and crowns were cut off. These could be used as food for stock, along with the valuable pulp left over after the sugar had been extracted. By this time the year was growing older and colder, and it was often in bleak winds, rain, frost or fog that we stood in fields turned to mud. With sacks tied around our waists, and hands covered in earth and numb with cold, we chopped away, a far cry from the sugar fields of sunny Jamaica. Gangs of Italian prisoners of war sent to farms to help with the harvest must have felt a long way from home, too.

After the beet was loaded into tumbrils and hauled to the roadside, it would remain in heaps until a permit came to take it to the railway station to be piled into high trucks, and taken to the factory. By this time the field had usually turned into a morass, with the journeys of the tumbrils creating a maze of ruts. The horses' hoofs sent great showers of mud across our backs as they struggled with loads through slush and ruts growing deeper by the day.

Since we entered the Common Market we have been compelled to produce sugar from beet—a subtle French revenge, perhaps, for that British blockade of long ago? But the production of sugar beet is different today—no more back-breaking hours of singling and hoeing, or of pulling, topping or loading beet. All the work is done from the seat of a tractor. Single seeds, precision drills and spaced seeding have done away with singling with a hoe. Sprays deal with the weeds, and any hoeing that is done is by tractor. Machines pull, top and load the beet. All that stays constant is the mud of a wet autumn.

To grow sugar beet, one must have a contract with the British Sugar Beet Corporation: without it you cannot even buy the seed. I do not think I shall try for one. I rather fancy that those labour-saving machines would too often be of little avail on our wet, heavy land—and I have no wish to return to a life of hand pulling, topping and loading.

# Old Tom

Old Tom was my father's carter; he had no interest in cattle, sheep or pigs, but he loved horses, and spent his life with them, working in the woods, tushing timber, hauling oak bark. Before coming to work for my father he had worked at the farm where I am now. One of his regular jobs was to take hay to the ponies in the Forest of Dean collieries.

This meant an early start in the morning; the horses had to be given their food and allowed time to eat it before the journey. As Tom's home was at Cinderford, he lodged in a farm cottage with the cowman and his wife. One night, though, the cowman was waiting for him outside the cottage and said: 'Thee cosn't come in yet, Tom, we've got visitors. Wait outside 'til they're gone.'

'Atter 'e spoke like thic to me,' Old Tom told me in later years, 'I packed my bag an' went away an' never stopped there no more.'

As a small boy, I used to be fascinated by the way Old Tom could drink cider. A pint mug would go to his incurved lips and down the cider went with no sign of a swallow. 'All cider's good,' Tom used to say, 'some's just better'n others.' But I don't suppose he tasted much, the way he drank it.

Tom used to boast that he could gain entry into most public houses, even when they were closed, and when I was older I had the opportunity to observe some of the wily tricks he employed. 'Once I'm inside, I'm all right,' Tom said. Nobody could recall seeing him drunk.

He was never at a loss, in any situation. And it was amazing what he could produce from his pockets. 'Half a tick,' he'd say, 'I think I've got just the thing in my pocket.' He never wore gumboots, always hobnailed boots—and corduroy trousers, tied just below the knee.

Old Tom never had any food before coming to work, but at about half past ten every morning, he'd say 'Ta-ra-ra bump de-ay, I've had no grub today,' and start to untie a bundle in a large red and white spotted handkerchief.

He was old when I first knew him, and he retired just before the war, but he grew bored and went to work as a blacksmith's mate in the woods. When the war ended he retired again—and grew bored again, and came back to work for my father with the farm's one remaining horse.

'It's good to be back with hosses,' said Old Tom. When he was ninety, he retired for the third and final time. 'I think I'll have a blow,' he said, just as he did when he rested his horses on the headland.

# Conversation on a shortening day

## October 1980

'Just a minute and I'll be with you,' said a disembodied voice as I walked into the garden; 'As soon as I've filled this basket.' That gave me a clue where to look and sure enough, there were two stout hobnailed boots on the ladder, a pair of corduroy trousers, and a bronzed arm stretching high for an apple in the tree.

Then with a squeak of a basket, the rattle of an iron hook and the clatter of boots on rungs, a man with a tattered straw hat atop a weatherbeaten face was beside me, thrusting a basket under my nose and asking if I'd ever seen such Bramleys.

'They're master Brambleys, ain't they?' he said.

I agreed they were indeed master Bramleys, and admired his large garden, most of it well stocked with vegetables and the rest neatly dug.

'Ah, I do every bit of it myself,' he replied. 'An' how old d'you think I am?'

'Sixty-three or four?'

A chuckle: 'I'm eighty. Never would have thought it would you? Eighty last birthday, eighty-one next.'

'And you do all this garden yourself?'

'Yes—*and* I spend a couple of days over there.' His arm pointed to some place distant. 'I reckon to work thirteen hours a day—but now it's getting as I can't manage it.'

So at last the years were beginning to tell on this spry old countryman? Not at all, for before I had had time to mutter some words of sympathy he had added: 'No, I can't manage it now the days are drawing in.'

I explained my visit: 'I've come for some spring cabbage plants.'

'Ah, I thought you had. I've got some toppin' plants,' he replied, but it was obviously not time yet for business: 'Y'know, I'm as bald as a billiard ball under this hat.'

He removed his hat with a flourish to reveal a full head of hair: 'Ha! That surprised you, didn't it?'

At last we came to the patch of cabbage plants. 'There, look at 'em,' he murmured. 'Beauties ain't they, kind and healthy, I think every seed grew. Now while I'm pullin' and countin' 'em you have a look round. Go over to that corner where the currants are. You'll see some master sprouts better'n those you saw way back.'

I walked along a grass path with trimmed edges until I came to the corner. The garden was a great deal larger than it had first appeared to be, sloping away at this point. Here again all was in fine order: soft fruit, more parsnips, leeks, carrots, swedes and sprouts.

On my return he'd finished pulling the plants and was straightening his back: 'Her don't half go on over, don't her? Some master sprouts down there too and all. You saw 'em, didn't you?'

I assured him I'd seen them, and agreed they were master sprouts. 'Like young trees,' I added for good measure.

'Nice bit of ground ain't it?' he said, scuffing the toe of his boot into the friable soil. 'But it weren't always like this. See over there.'

He pointed over the hedge to where I saw newly-ploughed furrows of stiff clay. 'It were like that when I came here, but I got working at it, dungin' it an' growin' mustard an' digging it in. I got at it an' gradually it come to; an' now it'll work easy an' grow anything I've a mind for.'

I paid him for the plants: 'You'll put 'em in tonight, won't you? An' give 'em a good soakin'.'

He seemed reluctant to give them to me, and only after I'd assured him for a second time that I'd take good care of them would he hand them over.

And then: 'Good day,' he called from the gate. 'I must get on. It's a job to get it all fitted in now the days grow short.'

# Hedges

Britain is still losing two and a half thousand miles of hedgerows a year, and in the past twenty-five years we have lost more than twenty million trees. Dutch elm disease and the effects of successive dry summers should have taught us that even trees do not last forever. Already we have fewer than any European country except Ireland, and now the Tree Council has declared the 6th to 12th of this month *National Tree Week*, to draw attention to the urgent need to plant more.

Trees are more than something good to look at; they provide shelter, shade and food for many kinds of wildlife; the oak alone attracts more than three hundred different species of insect. Oaks are a long time a-growing, and that is why we should plant them. Unless we plant oak we shall have no faith in the future, and until we have faith in the future we shall be miserable in the present.

The Tree Council hopes that everyone with the necessary space will plant trees, and during November many local authorities will be doing so. Celebrities will be planting trees; mind you, I haven't much opinion of the way celebrities plant trees, but perhaps, after everyone has gone, somebody will come back and plant them properly.

All this is admirable, but while I applaud such bodies as the Tree Council, and believe that grants towards tree-planting are public money well spent, I feel we need more than isolated areas of conservation. Isn't it absurd to be spending public money, too, on grants to encourage the grubbing of hedges?

The Ministry of Agriculture's regional publication *Phoenix* recently carried a short article stating that the destruction of trees, hedges, ponds and ditches had gone far enough; yet some Ministry officials are still persuading farmers to continue in this folly. I do not think farmers should be allowed to destroy trees or good hedges any more than beautiful and historic buildings should be allowed to be destroyed. Farmers, for their own good, should not be seen to be in a privileged position.

Trees and hedges are more than amenities and homes for wildlife; they also have an economic value as living things. Unfortunately, today, so much that does not bring a direct and almost immediate return is often dismissed as valueless.

The tree and the hedge provide shelter and shade for farmstock, sometimes of more value than the grass that could be grown in their place. Often, too, that seemingly haphazard pattern of fields and hedges is not haphazard at all. Usually a hedge follows the contours of the ground, and its ditches drain the land, or divide one soil type from another. Many a farmer must have made one big field out of five smaller ones and found that he still has five fields.

Without hedges, farming becomes more inflexible, and with the modern flail cutters their upkeep is comparatively cheap. Hindering machinery is a good point, but isn't this another instance of the machinery ruling us? And if we turn our countryside into a prairie, might we not lose, rather than gain, production in the long term?

Really, it is all a matter of balance, of common sense and reasonableness. We do not want any more regulations, rules and restrictions; better by far to have general consent. And there is a danger, I think, that some conservationists, through their zeal, could do more harm than good.

And this is where our Women's Institutes, among other organisations, could play a part. Now I know all about the jeers of 'jam and Jerusalem', but W.I.s are truly representative organisations. When it comes to the well-being of the countryside, they are always well to the fore, and their methods of gentle persuasion are unrivalled. Ladies—and many of you are farmers' wives: we need trees.

# The lovers' bus

'When we used to go in the colliers' bus to Cheltenham ...' said my friend from Cinderford.

'By colliers' bus to Cheltenham?' I asked, intrigued.

'Yes,' he replied, flatly. 'To see the girls.' His answer only deepened the mystery.

'The colliers' buses,' he began, leaving the most fascinating part to last, 'were really lorries for delivering coal. But with a top fitted, they were used, too, to take the miners to and from work, and then, about 1929, one of the owners started running one to Cheltenham to take the young men to see the girls.'

'But why go to Cheltenham?' I asked, unable to contain myself. 'What was wrong with the Forest girls?'

'There weren't any. No young unmarried women,' he answered, in that worldly but tolerant way that those a few years senior adopt. 'You wouldn't remember, but in those days there was virtually no female employment in the Forest, or not in East Dean. After they left school, many girls went to Cheltenham, into domestic service in the big houses, colleges and hotels.

'It wasn't until the last war that there was sufficient employment within the Forest, so you see, we, too, had to go to Cheltenham. Off we went on Sunday afternoons to meet our girls, or to find one, or just to look and hope to talk to girls. Perhaps someone had a sister we'd be able to meet, or maybe we'd sit on the Prom until they came by. Oh, they knew we'd be there, all right.

'At tea time they went back to work and later some would return, and if we were lucky they would bring us a slice of cake. After they'd all gone, we'd go and have a glass of beer before taking the bus back home.'

It seems incredible now, a whole district almost devoid of nubile women. Cheltenham then, of course, was very much the town of the retired gentry, and I can well picture these gauche working-class lads in their navy serge suits, loose white scarves, and flat caps—or, for the more dashing ones, bowler hats and dark coats.

There they would be, loitering or sitting beneath the Promenade trees, shyly eyeing the dawdling, giggling little groups of three or four girls. The bashful awkwardness of both sexes, the first few tentative approaches—this shyness of the youth seems almost unbelievable today, and it has a poignant charm. But many a Forest marriage came about this way.

# United under the old church roof

A few years ago the detached tower and spire needed extensive repairs; last year it was the organ, and now it is the roof of the church. There will be a grant, it is understood, but the parish will have to find £16,000 by today's estimate, and time and inflation will doubtless make the eventual cost much higher.

A formidable task—but like most country people, the parishioners of Westbury-on-Severn will strive to keep their church in repair, even if they do not attend its services very often.

Fund-raising began last August, during the bank holiday weekend, with the staging of *Village Life 1880–1980*. It opened with a scarecrow competition on the Friday evening; the judge was also dressed as a scarecrow, for he had been harvesting all day and was dirty and unshaven and late for his appointment. It was quicker and easier, he reasoned, to don a tattered smock and battered hat, to blacken an already dirty face and stuff smock, hat, trousers and boots with straw than to wash and change into respectable clothes.

Next morning the scarecrows lined a village street gay with bunting. Overnight a private house had been transformed into the Old Bakery; another had become the Old Sweet Shop, and at another coffee and doughnuts were being served, the china and tablecloths exquisite.

The previous evening the organisers had realised that they had a success on their hands, and there had been frantic late-night visits and telephone calls resulting in midnight baking and toffee-making.

Visitors came from afar; and motorists passing through stopped and visited the shops and stalls, inspected the exhibitions, and enjoyed lunches or cream teas at the farm.

A hundred years of fashion at one house; Victorian scenes at the school; kitchen, parlour and bedroom all furnished in the style of a hundred years ago. The couple in charge of all this Victoriana, when I called, could have passed as Victoria and Albert with a touch of make-up and the right clothes. I suggested as much, but their faces said: 'We are not amused.'

There were several exhibitions at the farm; old implements and tools, animals, a harvest scene, an old-fashioned dairy and much else. The goats kept escaping and eating the produce in the harvest exhibit, and this added greatly to the general liveliness.

The sun shone cheerfully, the church bells rang merrily, helpers, locals and visitors revelled in it all, and the roof fund benefited accordingly. Much thought and effort had been put into this village life weekend, yet the most important aspect of such life, and one that had been fast disappearing, sprang up of its own accord and blossomed wonderfully. The spirit of community returned to Westbury that day.

Though it was a happy occasion, to me, and I dare say to many other local people who have passed their half-century, it brought back bitter-sweet memories of joy and sadness. We remembered using many of those old tools and implements, museum pieces now. Some of those furnishings, clothes, tools and implements had belonged to relations and friends long since dead. And those photographs of bright young faces now grown old, and the smiling faces of those now no longer with us, reminded us of how things have changed even in our lifetime; 'They are not long, the days of wine and roses.'

And this, I think, is the reason why all of us, believers and non-believers, hold the church in such affection, and why we will rally to its repair when the need arises. We come and we go, as shadows fleeting over the landscape, but the church remains, giving us a sense of perspective and continuity; linking us with those who have gone, and those still to come; and extending a glow of comfort beyond the confines of dogma; mysterious, unspoken, dimly-comprehended perhaps, yet the more real and powerful because of that, whatever our beliefs may be.

# Swept on by the Severn

## November 1983

It was a good bright morning, and the tide was running well when my wife and I boarded *The Riparian*. The Severn fascinates me, summer or winter, high tide or low, day or night—especially at night, when the moon is shining on her—but I am frightened of the water.

My wife loves it; in fact she has it in her blood, for her mother's father was a sailor. A good many years ago, she and two of the boys lured me in to a little boat on the Thames at Henley. She was a little boat, too, and with four of us she was overloaded; I could see that quite plainly, but they couldn't. I could also see a very large boat—a huge one—coming towards us, and I could plainly see the danger we faced from her wash.

They laughed at my fears, but that was only because they couldn't see what I saw only too clearly. Nothing untoward happened to us, but it was only by luck we didn't all end up in the Thames, as I still tell them. I have never liked the Thames since.

But *The Riparian* was a big boat, one used for carrying stone, and I had every confidence in her skipper. 'We'll let her drift for a bit,' he said when he cast off—if that is the correct term. We did indeed drift with the tide, quite happily, and it was only when he said: 'I'll have to start the engine or we'll be into the cliff' that sure enough, I could see us heading straight for the obstacle in question.

Why didn't he hurry up and start the engine? He'd be into the cliff in a minute. I'd half a mind to tell him so, but I kept quiet. He went down the ladder, and though I could hear his tinkering about, there was no sound of an engine coming to life. Perhaps it wouldn't start—and there we were, sweeping closer and closer to the cliff.

Then the engine started; and not a moment too soon, and I was glad I hadn't voiced my fears. You see, I had my sailor's cap on; no matter that it also serves as my working cap, my going-to-the-pub cap, my general all-round cap. This morning it was a sailor's cap. That, I believe, is what it was made for, and on no account must I disgrace it.

We chugged merrily but sedately up the Severn, and admired Newnham Church on the cliff top. Newnham has lost several churches, and viewed from this angle it looks in danger of losing another. Ah, I thought, they throw the trimmings of the churchyard over the railings and down the cliff. Someone cuts the grass at Newnham Ferry, but someone else throws all manner of rubbish over the cliff. A pity the one action spoils the effect of the other, but that's the way of the world, I suppose.

Newnham looks nice from the land, but from the river it looks romantic, with the view up Severn Street, the old George's timbers at the top, and the warehouses at the disused quay. It is not parochialism that makes me describe only my side of the river, but we had to keep to this side because of the sand on the other. Once the sand was this side, and then the boats had to wait to dock over here. The pub across the river was a favourite haunt, and a fair old rowdy place it could become, by all accounts.

Pillboxes were built on the other side just before the last war. All of them were several yards inland then, but all except one have tumbled into the river through erosion.

How different everything looks from the river, past the fish house, the hills and woods beyond; Broadoak; the White Hart; the old school. I see where some of Broadoak's rubbish goes, and on we go, past Mr Jackson's landing stage, past Spanker's; I never knew Spanker had all those greenhouses.

Now we come to the place where erosion is biting seriously into the land. See the home-made efforts one man has made to save the remainder of his garden. Unless he succeeds he will lose his house, too, judging by the amount of ground I know he has lost in recent years. The river board's wall to prevent erosion stops some way above his property, so he has been forced to do what he can in his own limited fashion, and at his own expense. It seems unfair, for if the river board is in charge of the river ...

Like all official bodies, I expect it has its own mysterious regulations and restrictions, which fail utterly to regulate or restrict the river. But my space has come to an end, now, far faster than our river trip did. The trouble is, I will digress. I should have kept moving, like the Severn tide and *The Riparian*.

# Warren James, reluctant rioter

## November 1986

On Wednesday June 8, 1831, a group of local people under the leadership of Warren James began to demolish the substantial fences which had been put up some 20 years previously to enclose the Forest of Dean. The enclosures had deprived Foresters' animals of the grazing to which they had become accustomed during the long period when management had been slack, and the rioters believed that right was firmly on their side. They contended that the fences had been kept up longer than was necessary for the young trees and that more than the permitted 1,000 acres had been enclosed, and they gave due warning of their intentions that Wednesday morning.

Two days earlier Edward Machen, deputy surveyor of Dean, had notices displayed to the effect that the enclosures were lawful and that to lay them open was illegal. He had also tried to reason with James about his proposed action, to which he replied that 'the Forest was given to us in Parliament last year.'

Warren James was a freeminer who lived at Bream. He was of good character, mild mannered, reserved and peaceable, and had never done anything without much deliberation. Abstemious in his habits and a regular attender at Parkend Church, he must have been convinced that he knew the reasons for the wretched conditions of the Foresters, and felt entirely justified in his course of action. He also believed that he had the support of people of authority in London.

The Forest at this time was also feeling the effects of the Industrial Revolution. Foreigners—those not born in Dean—were taking over the work which had once been the exclusive right of the freeminers.

Here the freeminers were at a disadvantage, and not only because the foreigners were wealthy and seemed to have support from the Crown; the documents of the Mine Law Court, which would have afforded them protection, had been stolen in 1777. Sir William Guise, a Gloucestershire MP, reflected after the riots that their basic cause had been the abolition of the Mine Law Court, and he was prominent in urging the Government to re-establish it.

The Forest disturbances of 1831 stemmed from much the same causes as the more famous Swing Riots of the same period. The Foresters were suffering from loss of liberty and their means of livelihood, from hardship, poverty and starvation. The riots, their cause and aftermath are an important part of Forest and indeed national history, and they have not in the past received the attention they deserve.

That omission has now been remedied with the publication of *Warren James and the Forest Riots* by Ralph Anstis. Here we have a full and engrossing account, well researched and documented and with the Foresters and their plight treated with understanding and sympathy.

When the uprisings began the authorities did little except read the Riot Act, watch and confer. Their inaction served to encourage the protesters in their belief that their actions were right, and that the Crown knew it. Eventually the Militia was summoned from Monmouth, twice it was marched up the hill to Coleford, and twice it marched back down again. Coleford people, whose hearts were with the rioters, laughed and called it 'the ragged army.'

By then the rioters were a force of hundreds, including women and children. Authority, greatly outnumbered, was prudent and did nothing. Within a few days 60 miles of fencing was destroyed, but there was no violence to person or property, and none of the trees was damaged. Edward Machen, the deputy surveyor, behaved in exemplary fashion throughout the affair. Before, during and after the riots he showed tolerance and kindness, even writing letters and signing petitions in support of rioters, and there is little doubt that the subsequent leniency towards the Foresters was largely due to him. He almost bids to rival James as the hero of this book.

Warren James, however, was not treated leniently, and when it comes to his trial authority is not seen in such a good light. He was sentenced to death, a maximum penalty secured on inadequate evidence, but was later reprieved and transported for life. The same fate befell John Harris.

Mr Anstis gives graphic descriptions of conditions on the convict hulks and on Van Dieman's Land—now Tasmania—where James served his time. In 1836 he was pardoned but he never returned to England, since he had no money and was sick. He died in Hobart in October 1841, aged 49.

Just over ten years previously he had fought with all his might for his fellow Foresters—'Foresters', by the way, with a capital 'F'. One of my few quibbles with Mr Anstis is his constant references to 'foresters', which gives the impression that all inhabitants were men of the trees. Had James not made his stand, it is doubtful whether the freemining rights of Dean would still be with us today.

The author makes interesting speculations about that unknown London support, the hustle to spirit James way from England and his eventual pardon, with pertinent remarks throughout. This is a book for all true Foresters and many more as well, a moving story told in a compelling manner, and I hope it receives the wide readership it deserves.

# Bullo Docks

November 1980

Some two miles south of Newnham, just off the A48, is the tidal creek of Bullo Pill. An Elizabethan man-of-war was built there, as well as other ships, and many more were launched from Newnham, where Henry II and his army met Strongbow, Earl of Pembroke, before they embarked for Ireland.

The port of Newnham lost out to Bullo Pill with the coming of the railways. By 1800 Bullo's shipbuilding had declined, but seven years later John Rennie, a civil engineer, was asked to report on Forest of Dean railway schemes, and he saw the pill as capacious and conveniently approached; just the place, he suggested, for a little dock.

A harbour and quays were built between 1808 and 1818, and a mineral line from the Forest opened in 1809 included a railway tunnel, the first in Britain—and possibly the world. Coal was offered for sale from Bullo Pill in June, 1810, and came to Gloucester three months later via Bullo Dock, the new tramroad and the tunnel through Hay Hill.

In 1814 there was a contract to deliver three hundred tons of large coal daily to Bullo Wharf, at eight shillings per ton. By 1815 Bullo could ship a thousand tons of coal and stone daily, and these along with timber, bark, pig iron, slate and hides from nearby tanneries, were shipped from here regularly in the nineteenth century. By the middle of the century, even Lydney was losing trade to Bullo.

The Gloucester to Chepstow Railway was opened in 1851. Four years later the Bullo track was widened to take locomotives and linked to the South Wales line, only to be converted to the narrow standard gauge in 1873. A locomotive shed, passenger halt and a goods station followed, in time—to be closed, respectively, in 1931, 1958 and 1963. More recently the railway bridge spanning the A48 has been demolished.

A small industrial centre grew up around the dock: a marble works in 1824, the Standard Wagon Company in 1897 and the Bullo Docks Concrete Company in the 1920s. The wagon company was replaced by Healey Brothers, makers of perambulator tyres, and is now the Newnham Rubber Mills.

In 1872 sixty-four craft were trading regularly to Bullo; sloops, trews and barges. But by the turn of the century, 1903, trade had declined to only eight cargoes a month, and by the 1920s there was little more than the occasional trow. Appropriately, the barge *Finis* was the last to use the dock on business, in 1926; shortly before the Second World War hydroplane racing brought Bullo Docks a little brief fame, but that aside, their rise and decline were linked inexorably with the fortunes of the river and the railways.

Today the small dock is silted up with mud, its gates in ruins and their oaken timbers slimy and decayed. The coal chutes, drawbridge, tips, cranes and railway lines are wrecked or gone, and the ponds overgrown.

The riverside quay remains solid and impressive, and birds-foot trefoil grows among the shortcropped grass that was once the scene of so much hurried activity, with cargoes loaded when the tide was right and every able man, woman and child was pressed into service.

Discarded timber, rusting ironwork, crumbling stonework and rank weeds add to the desolation; yet there is a fascination about Bullo Docks even in decay.

See them on a dull November day; the vast expanse of Severn, mud and sand; the trickle of the stream at low tide, as the water edges its way through the mud of the dock; the forlorn cry of a seagull; and in the distance Newnham Church, high on its cliff, silhouetted against the grey autumn sky.

Bullo Pill failed as a centre of industry; but the scene it presents today is fit for any poet or artist.

# My fuddled friends

## November 1978

We'd been going out together for two or three months before she said anything about it. The time was thirty years ago almost to the week, and we were courting. She lived on one side of the Severn, I on the other. Her mother had told her when she was a little girl that our side was where the wild men lived. We often met in Gloucester; occasionally we would venture further afield.

And then one evening she observed: 'You have a lot of friends.' It seemed a strange remark. 'Yes,' she continued. 'Wherever we go we meet your friends, and almost all of them are drunk.'

Ah, I knew what she was talking about now.

'But they're not my friends,' I said.

Now *she* was puzzled: 'But they come up to you in the street; cross the road, even hurry up to you to greet you effusively; shake your hand; slap you on the back; are as friendly as can be; smile and laugh, and are obviously delighted to see you; are equally obviously drunk, and are loth to leave you.'

It was all quite true, though I'd grown so accustomed to it I scarcely gave it a thought. But as I explained to her, they were not my friends; I'd never seen them before. She didn't believe me: 'But you talk to them as though they're dear old friends; they and you seem overjoyed at the meeting. They even seem to recognise you in the dusk, and come staggering along to meet you.'

Yes, she was quite right. For some reason I'd never been able to discover, I had an irresistible attraction to drunks. Almost any drunk within four or five hundred yards would make for me. It couldn't be because I had a kind, sympathetic face—I may have had, but no one ever told me so.

No, until further information came, I had to accept that to drunks I was a magnet. 'Yes,' she replied hesitantly, and probably remembering what her mother had told her years ago, 'but how do you explain your friendliness, your delight, and the fact that you know what to say to them to give *them* such delight?'

That was easily explained, if not so easily believed; what I didn't know by instinct, I'd learned by experience—and, by God, I'd had some experience. Quite simply, it was the only way; any other way and these jolly, friendly gentlemen soon became very ugly customers indeed.

Well, she accepted it, but I could tell she didn't like it, and ever afterward she kept a wary eye open for any of these merry, staggering fellows. And at the slightest hint of one's proximity, she'd try to hustle me away, usually without success; they'd follow hard on my heels shouting: 'Hey, hey, just a minute!'

The only thing that I could do was to stop and chat with the dear fellows—I didn't want to cause offence—and she'd wander on thirty yards. She never said much about it, but I knew she didn't like it.

Then we were married and didn't go out as much, but I still had occasional encounters, sometimes during the day. There was the time when our children were small, and I met a particularly jolly chap, who, despite his stagger, simply refused to part from me.

He was very fuddled, but too cunning to be easily lost, following us from shop to shop, running along side streets, most earnestly desiring more of my company. I could tell that my children liked it no more than my wife.

There were occasional incidents during the following years, until about three years ago. And then no more; perhaps drunkenness wasn't so prevalent, or I wasn't about in the right places—they must be chance encounters in the street. Or perhaps I'd simply lost the power.

And then, a few weeks ago, at four-thirty in the afternoon in Cambridge: oh, he was so unsteady and jolly and friendly, he smiled and winked and was quite delighted to see me; the street was crowded, my wife grasped me by the arm, and propelled me on at speed.

He was too dumbfounded at my sudden departure, too unsteady to follow. He just stood there, utter dismay on his face, oblivious to all the others, crying: 'Don't go, don't go, stop just a minute.' My wife hustled me on regardless.

I was sorry to leave him, poor friendly fellow: but simply delighted that I hadn't lost the power.

# Christmas is for ever

Old December's bareness everywhere; the beginning of this month is the beginning of winter, even if the calendar would have us believe otherwise. By now we are all too familiar with the leafless trees and hedges, and the first hard frosts, and December's austere landscapes and grey skies can even have their charms, so long as there is not rain day after day; rain means mud, and mud is the farmer's winter enemy.

First in the gateways, it advances into the fields, and towards house and buildings. It seeps into our souls. Cattle on pastures turn the fields into a morass scarcely recovered by half a summer's sun. Cows still on kale struggle in it and become plastered, their udders sore and their feet lame; to clean them for milking is a demanding task.

Tractors spin helplessly in the mud of their own making. It is then that, like King Richard, I cry vainly for the horses we abandoned years ago, because they were too slow; hardly as slow, though, on reflection, as a tractor axle-deep.

Mud, rather than the calendar, determines when we house the cows. I still keep cattle in covered yards, unlike many dairy farmers who now use cubicles or cow kennels, a system which only increases another winter problem, slurry—wet, sloppy muck, which, even when spread on the land, is no match for our mixture of dung and straw.

It is a relief when the cattle are safely under cover. I feel a sense of well-being to see them comfortably housed, with dry backs, to hear the rustle of straw and the steady munching of hay.

On any livestock farm it is almost impossible to have sufficient covering for all the animals, produce and machinery. During the winter, every space is crammed. To build another shed is simply addictive; all too soon, we are wishing, we had just one more again.

Once, during the dull days, the farmyard was alive with the gobbling of turkeys, the cackle of geese, the crowing of cockerels. And then, round about the time of the winter solstice, there was a silence, and soon fingers were sore from plucking. Now few farmers' wives fatten poultry for Christmas; like butter- and cheese-making, the trade has moved from the ordinary farm.

It is easy to become nostalgic at this time of year, and yearn for the good old-fashioned English Christmas. Mr Pickwick's visit to Mr Wardle's farm at Dingley Dell is idyllic, the archetypal good old country Christmas, but far more people have far better Christmases today than then. And how much of the popular idea of an old English Christmas, in fact, is either English or indeed very old?

Snow at Christmas is rare in England, and if a heavy fall prevented the reuniting of families and friends, it would hardly be welcome, anyway. Yet I do see one affinity of snow and Christmas; both have the power to make us nicer to our neighbours.

Carols, now an integral feature of Christmas, were ignored by the Church and the wealthy for two hundred years, and survived only because the humble never forgot them. The Christmas tree was unknown here until some German families in Manchester introduced it in the nineteenth century, and later the Prince Consort popularised the custom.

Plum puddings contain ingredients from all over the world, but not, as far as I am aware, our English plum—perhaps it was originally plump pudding, anyway. And boar's head, roasted swans, peacocks and bustards—though not quite the goose—have been replaced by the Mexican turkey.

Kissing under the mistletoe, at least, is authentic and English, although now enjoyed all over the world. Like the evergreen, the symbol of undying life, it has survived from before the days of Christianity, as, of course, has this winter festival itself.

Christmas Day on the farm starts earlier than usual. Perhaps there will be stars overhead as we bring the cows in to be milked, for Christmas is a time for stars, rather than mud.

Black shapes in the gloom of the covered yard grunt and clamber to their feet. Only as they loom towards the milking parlour do they take on their familiar form, and become once again the Daisy, Nutmeg, Foxglove and Rosebud of old. A happy Christmas to you all, girls!

# Christmas cards—and cads—
of long ago

December 1980

'Happy, happy Christmas that can win back the delusions of our childish days …' At this time of year, what better excuse do I need to write about *The Magnet*?

*The Magnet* was published every Saturday. In my time it cost 2*d* and had a pink cover, and today all thoughts of it, particularly the illustrations by C.M. Chapman, fill me with nostalgia and delight.

Then there were those tiny advertisements. They intrigued me more than forty years ago, and they intrigue me still: 'Your height increased in 12 days or no cost.' 'Have you a red nose? Send a stamp and you will learn to rid yourself of such a terrible affliction free of charge.' 'Blushing. Free to all sufferers, particulars of a proved home treatment.'

Started in 1908, *The Magnet* was devoted to the saga of Greyfriars, a minor public school somewhere in Kent. Until the paper folded thirty-two years later, neither the boys nor masters aged. 'The year is 1910 or 1940, but it is all the same,' George Orwell wrote critically.

Christmases came and, free from the gimlet eyes of old Quelch, the boys of the Remove went off for the hols: Harry Wharton, Bob Cherry—I never really cared for him, with his breezy 'Hallo, hallo, hallo!'—Nugent, Bull, and Hurree Jamset Ram Singh.

Vernon-Smith, the bounder of the Remove (and my favourite character) usually accompanied them, with the gross Bunter tagging along.

Bunter became the star of *The Magnet*, but one thing that puzzled me about 'the tightest trousers in Greyfriars' was why he was allowed to wear those loud checked trousers in the first place. The Fat Owl of the Remove had 'paws', never hands, and 'rolled' rather than walked. He spent his lengthy childhood always expecting a postal order, stealing other fellows' tuck, lying, being found out, and kicked: 'Oh, I say, you fellows. Leggo! Oh crikey! Urgh! Yarroo!'

Nobody wanted Bunter at Christmas, but Christmas would not have been the same without him; the festivities were spent at Wharton Lodge, Mauleverer Towers or some moated manor house, all of them deep in snow and wrapped in mystery.

My best-loved Christmas present in those days was the fat *Greyfriars Holiday Annual*, which also contained stories of St Jim's and Rookwood schools. The St Jim's stories appeared weekly in *The Gem*, and so for a time did the story of Frank Richards's schooldays, both written by Martin Clifford.

Owen Conquest wrote the Rookwood stories and Frank Richards the Greyfriar ones; but Clifford, Conquest and Richards, they were all the same man, Charles Hamilton, who had a score or more further pseudonyms. When he died on Christmas Eve, 1961, it was estimated that he had written the equivalent of a thousand full-length novels.

His style was repetitive and cliché ridden; proverbs and quotations abounded—and no wonder, with two full papers to write each week. 'He! He! He!' 'Ow!' 'Yow!' 'Ugh! Ugh! Ugh!' repeatedly padded out the dialogue. His writing was unique and inimitable; he was the master of his craft, and gave pleasure to generations of schoolboys.

In 1940 George Orwell wrote a slashing attack on Greyfriars and Richards in *Horizon*. Richards later counter-attacked with verve and vigour in the same magazine, but time was running out for him. Soon afterwards the wartime paper shortage killed *The Magnet*, having already accounted for *The Gem*.

In spite of his prodigious output, Richards was not a rich man, and what money he had earned was lost on his pre-war visits to Monte Carlo. Forbidden by his contract to continue the Greyfriars and St Jim's stories elsewhere, he was reduced to penury and obscurity. Later he wrote school stories in other papers, and Billy Bunter appeared in hardback, in comic strips, on the stage and on television; but the old magic and charm of *The Magnet* days were never recaptured.

Since his death, the nostalgia industry has ensured his reputation. Old *Magnets* were republished in bound form, and scores of the volumes are now in print.

The oak trees at Greyfriars have shed their leaves; there is a nip of frost in the air. The skies are grey, and heavy with the promise of snow.

The quad resounds to laughter, even old Quelch allows himself a smile, and the Christmas holidays are beckoning Harry Wharton and Co. to their moated manor house. Bunter has not been invited, but Bunter will be there ...

It is 1910, or 1940, or 1980, it is all the same. Of course it is; these fellows are immortal.

# My yearly escape
# on the Muggleton coach

Christmas is a time to renew old friendships—and as it happens, it was at Christmas that I first made the acquaintance of some of my oldest friends. Forty years ago I bought a copy of *Pickwick Papers*, and almost every Christmas since I have revisited Mr Pickwick and company, turning inevitably to the section which Dickens called 'a good-humoured Christmas chapter.'

With what delight we board the Muggleton coach with Mr Pickwick, Mr Snodgrass, Mr Tupman, Mr Winkle, Sam Weller, the huge cod-fish and the six barrels of native oysters. The horses speed at a smart gallop, tossing their heads and rattling their harness, and the wheels of the coach skim over the hard and frosty ground.

At three o'clock the company arrives at Dingley Dell, high and dry, safe and sound, hale and hearty on the steps of the Blue Lion. Mr Pickwick feels his coat being pulled, and 'Aha!' 'Aha!' says the fat boy, Mr Wardle's favourite page.

Messrs Pickwick, Snodgrass, Tupman and Winkle set off across the fields to Manor Farm, the home of their jovial friend Mr Wardle, leaving Sam and the fat boy to attend to the luggage. 'Vell, young twenty stun,' says Sam. 'You're a nice specimen of a prize boy, you are.'

The Pickwickians are greeted by a loud 'Hurrah' from Mr Wardle; and we are back in the eternal, idyllic world of a Dickensian Christmas once again. On Christmas morning Mr Pickwick meets Mr Benjamin Allen and Mr Bob Sawyer, both medical students.

They have been smoking cigars and drinking brandy, the one with his legs upon the kitchen table, the other with a barrel of oysters between his knees. Coarse fellows, the sort who shout and scream in the street, but Mr Pickwick excuses their habits as 'eccentricities of genius.' Eventually, Bob Sawyer has a medical practice in Bristol.

There is the journey from Bristol with Mr Pickwick and Allen inside the chaise, Sam Weller in the dicky, and Sawyer on the roof with legs asunder, a case bottle of milk punch and an enormous sandwich, howling and exchanging pleasantries with anyone he sees, a red flag of his flying from the dicky. No wonder the people of Gloucestershire stare; at Berkeley they halt at the Bell for lunch, and at the Hop Pole at Tewkesbury they stop to dine.

I still have that book of forty years ago, and have since acquired the first published volume of *Pickwick Papers*, but the Pickwickians' first appearance was in a series of monthly instalments. The publishers wanted someone to supply the text for a series of sporting adventures to be illustrated by Robert Seymour. Eventually they approached Charles Dickens; he was only twenty-four and virtually unknown at that time, but he was not prepared to play second fiddle, and agreed only on condition that Seymour should illustrate his story, and not vice-versa.

At a later date, Dickens was adamant that Pickwick was his invention, not Seymour's, though I suspect the original idea was conceived by the publishers, and owed something to Surtees's Mr Jorrocks. Anyway, the sporting aspect was soon forgotten, and as the story progressed, Mr Pickwick changed from a rather pompous type into a wise, kindly old gentleman.

The first number of the series was issued in April, 1836. Seymour died after only seven illustrations, having defined Mr Pickwick with his bald head, tiny spectacles, tights, tails and gaiters, and Hablot Knight Browne, better known as 'Phiz', replaced him. Phiz had illustrated Jorrocks, and was to illustrate many of the Dickens novels.

The early numbers were not particularly successful—and then Sam Weller appeared. Sam was boot cleaner at the White Hart: 'Number twenty-two wants his boots.' 'Ask number twenty-two vether he'll have 'em now, or vait till he gets 'em,' replied Sam. When he became Mr Pickwick's manservant, the series grew into a huge success.

*Pickwick Papers* is a vast, rambling, humorous book, an antidote to our present troubles. It is full of comic situations and crowded with comic characters—about a hundred of them.

The best and the greatest of them is Samuel Weller: but Sam is much more than a comic character; he is the personification of the best of the English character, good-natured, witty, cheerful, resourceful, kindly and loyal. I would rather have Sam Weller's company and friendship than that of a score of more exalted folk.

# When Christmas was as hard as winking

## December 1983

Christmas comes but once a year—but commercial Christmas comes earlier year by year. Before the leaves of autumn have finished falling, those asinine TV jingles assault the ear and insult the senses. Then the shops sprout synthetic holly and frost, the streets are crowded and the cash registers jangle, for it seems that everybody but me can find just the right things to buy for presents. I like Christmas by about four o' clock on Christmas Eve, but how I hate the weeks preceding it, except when I am in the fields.

Christmas comes sooner to all of us as we grow older, and with age come memories. I can still recall the thrill of finding the filled stocking, and Christmas—indeed, life itself—was not the same when Father Christmas no longer visited me.

Then there were the parties. The teas always started with bread and butter and jelly, which I disliked, but even more I loathed the games: musical chairs, postman's knock, and worst of all the one in which you had to wink. I was unable to wink, and unable to be jolly to order, for that matter. They are two disabilities I have never outgrown.

Even now, if I go to a party, I invariably find myself alone and silent in a corner while all around I see people laughing and talking. There must be something the matter with you, I tell myself, yet normally I am garrulous, or so I am told.

Christmas morning forty years ago, just left school, sitting on a three-legged stool, bucket clasped between the knees, head thrust

into warm flank of cow. 'Ping' went the first drawn milk as it hit the
bottom of the bucket, and later as it began to fill, the sound changed
to a gentle 'fuzz-fuzz'. There were other sounds in that lantern-lit
shed; the jingle of the chains around the cows' necks, the rustle of
straw beneath their feet and their steady munching. The air was
redolent of cows and the pulped swedes they were eating.

We pulped those swedes by hand, turning a handle round and
round; they were hard to pulp, too, much harder than mangolds,
which were never used until after Christmas. After milking there
were more swedes to pulp for the bullocks in the yard. Never mind
how cold the morning; ten or twenty minutes on the pulper acted as
a wonderful warmer.

The pulped swedes were carried in a large oval basket to the yards
where the fat bullocks stood waiting, issuing from their nostrils
twin jets of breath in the cold air. All their drinking water had to be
carried in buckets, and on frosty days they always seemed thirstier.
Carrying water was a slow, laborious job, and with each successive
journey our hands grew colder. Cutting hay out of the rick with the
big salmon-shaped knife was a pleasant task; the cut hay smelled of
summer and I felt it retained its warmth, too.

There were so many handles in those days, pulper handles,
milk- and water-bucket handles, hay-knife handles, pike and fork
handles and shovel handles. A handle for almost every job, and
every job done by hand. Now I wonder how we did it. Today all
that hand work would be accounted drudgery; perhaps it was,
but to most of the work then—except water-carrying—there was
a rhythm. Usually there was company, too, and we laughed and
talked as we worked.

Ten, twelve, fourteen years on; the morning air at Christmas
was redolent of silage instead of swedes. A milking machine had
replaced the bucket and stool, neither cows nor bullocks had
horns, and a pocket-knife had replaced the hay-knife because
hay was tied up in parcels; but water still had to be carried by
hand—and Father Christmas had come again. I relived the same
old thrill as my children shouted with delight at the feel of filled
stockings.

And so on to 1983; the Christmas morning air will still be redolent
of silage, but water no longer has to be carried. Milking done, cattle
fed, breakfast eaten, logs on the fire; this, as Mr Pickwick observed,

is real comfort, as I sit and wait for the arrival of my grandchildren. Indeed it is more than comfort, it is pure delight, for Father Christmas is alive and well again.

# The strange tale
## of the summery Santa ...

### December 1984

One hot afternoon last summer, my wife and I were in Newnham with our grandchildren, Harley and Carlaina. We were in the broad High Street, close to the bank, and it was such an ordinary afternoon until Harley pointed to the bank and shouted: 'Look, there's Father Christmas!'

We looked and there, coming out of the bank, was a portly gentleman with a white beard and a benevolent face. 'Father Christmas,' said Harley, in awe and eyes astare. 'Father Christmas,' said Carlaina with eyes wide open, but in a very matter of fact tone of voice, as though the appearance of Father Christmas in broad daylight—and in the height of summer, too—was quite a natural phenomenon.

He certainly *looked* like Father Christmas as far as his face and beard were concerned, but I was not convinced that he *was* Father Christmas. Harley and Carlaina talk about him the year round, but to see him in summer; well, it was rather more than I could accept, especially as he was in shirt sleeves.

I didn't wish to spoil their obvious enjoyment, yet I felt compelled to raise some objection. 'He's not wearing his red coat,' I said.

'Of course not,' replied Harley scornfully. 'It's too hot.'

'Father Christmas,' repeated Carlaina firmly, to quell my doubts—and both would have rushed along the street to greet him, had they not been restrained by Grandma. Perhaps she should have let them go, but she, like me, still had her reservations. Yes, he certainly *looked* like Father Christmas, and Harley had given a satisfactory explanation for the absence of the red coat; but even benevolent gentlemen with white beards who look like Father Christmas may not care to be greeted as Father Christmas if they are not, in fact, Father Christmas.

The gentleman spoke to a lady who had been standing outside the bank, and the two of them walked slowly up the street. 'That's Mrs Christmas,' said Harley. 'Mrs Christmas,' said Carlaina, as though that settled all doubts, all argument.

The lady and gentleman stopped by a motor car parked in the street, and after talking for a second or two, they both climbed into it.

'I didn't think,' said Harley in a surprised tone, 'that Father Christmas would have a motor car.' This was evidently a shock; he just stood and stared, and even Carlaina was silent. In fact all four of us stood and stared in silence as the motor car drove away.

When it was out of sight, Harley said: 'I never knew he had a motor car.'

'Father and Mrs Christmas gone,' was all Carlaina could muster.

All this was so unexpected, so extraordinary to me, so seemingly normal and believable to Harley and Carlaina, that I could not help asking a few questions.

'I thought Father Christmas only came at Christmas.'

'He only brings presents at Christmas,' said Harley. 'But he's real all the time.'

'I thought he lived at the North Pole or somewhere where there's always snow,' said Grandma.

'Not all the time,' explained Harley. 'I expect he gets tired of the cold all the time.'

Well, that seemed reasonable. He had to be somewhere, and why *not* Newnham? It is, after all, as nice a place as anywhere to be in summer—or winter, for that matter; a great deal nicer than a lot of places.

'What,' I asked, 'was he doing at the bank?' I thought that would stump Harley.

'He'd gone to get money to buy stuff to make toys for us at Christmas,' he answered.

'Toys, nice toys, and sweeties,' said Carlaina, taking her cue, as always, from her elder brother.

When Harley and I visited the bank a week later we made inquiries. But you know how close banks can be about their clients, and the answers we received were very noncommittal.

Perhaps it was because of this, or perhaps it was because of further thoughts about that motor car, but Harley, too, is beginning to express doubts. Now winter is here Father Christmas in shirt sleeves does not seem so believable, maybe, especially as the approach of Christmas brings the realisation that the real Santa is mysterious, elusive—and never seen.

# Our dear departed line

December 1977

Life hasn't been the same since they killed our railway. We all told the time by the trains, got up when we heard the first one, went to lunch, or dinner, when we heard the twelve-forty-five coming from Gloucester.

If you travelled on this train, all along the line you'd see men in the fields, laying down their tools and donning their jackets; in those days you could see men in the fields. The three-twenty, on the way to Gloucester, meant it was about time to fetch the cows for afternoon milking; the five-twenty meant tea-time. And the last train of the day told us it was time for bed.

We were able to forecast the weather, too, by the sound of the trains. The Gloucester-Hereford line ran alongside our fields on the north side, and if the trains sounded loud and clear, we knew the weather would be fine. The clear sound of trains on the South Wales line, a mile to our South, foretold rain. Strangely, now that the Gloucester-Hereford line is no more, the trains on the other line tell us nothing.

When I travelled by train to market, or my wife for shopping, we met our neighbours; now we rarely meet them. Today, whether we like it or not, country people are forced to have motor-cars; the increasing cost of motoring bears heavily on country dwellers, to whom motors have become a necessity. If all the oil fields should gradually dry up, I can't help thinking the gains would outweigh the losses. And if it meant no more aeroplanes, the world would become a safer and a larger place.

Our Gloucester-Hereford line was opened on 1 June, 1855, a broad-gauge, single track running from Hereford to Grange Court, where it joined the South Wales line. It was twenty-two-and-a-half miles long, with four viaducts over the River Wye and with tunnels at Lea, Fawley, Ballingham, and Dinedor. At first the line was leased to the Great Western Railway, but in 1862 it took it over completely.

The Gloucester-Hereford was the first line that the GWR converted to narrow-gauge. Three hundred plate-layers were employed, and buses were hired from London to take passengers from Hereford to Ross, and from Ross to Gloucester, while the work was under way. Work began on Sunday, 15 August, 1869, and was expected to take about fourteen days. In fact the conversion, including one-and-a-half miles of tunnelling, was completed in just five, and on Friday, 20 August, the first narrow-gauge passenger trains were running between Hereford and Gloucester.

Our station, Blaisdon Halt, was opened at a much later date. It was in the Westbury-on-Severn parish, which once had the distinction of having three stations: Blaisdon, Westbury Halt, and Grange Court. Today it has none.

I have fond memories of loading or unloading cattle, corn, plums, sugar-beet, feeding-stuffs and fertilisers at Blaisdon. Unlike the unpredictable and impatient road transport of today—lorries always seem to arrive at awkward times—the railway allowed us to load or unload when it was convenient to us.

A one-time porter at Blaisdon, born at Westbury, achieved considerable distinction. He moved to Ross, and, while still a porter, became the president of the National Union of Railwaymen in 1941. During his three-year term of office his abilities were noted by the Prime Minister, Winston Churchill, and instead of returning to our line he served on a commission on constitutional reform in Ceylon, and a committee on legal aid. In 1945, the new Prime Minister, Clement Attlee, appointed him Governor of Bengal; and Fred became Sir Frederick Burrows.

It was in Calcutta that he made that celebrated remark about knowing nothing about hunting and shooting, but a lot about shunting and hooting. A few years before his death, when he still lived in the same house at Ross that he had occupied as a porter, he made a valiant attempt to save his line—our line—from extinction.

Once more in its history the line had to call upon motor transport. On 28 March, at the end of that terrible winter of 1947, the Wye was in flood and the centre pillars of the bridge between Backney and Fawley were swept away, leaving almost a hundred feet of track without support. Luckily, the last train of the day had passed, and disaster was averted.

Sherlock Holmes and Dr Watson travelled this line when they investigated the Boscombe Valley mystery. Possibly the great detective was too preoccupied to notice the countryside, but I'm sure Dr Watson must have admired it. The dramatic changes of scene; hills and valleys, orchards and woodlands, fields and farmsteads, the red arable land of Herefordshire.

The line remained 'steam' until the end, and the last passenger train ran on 31 October, 1964. A few goods trains continued for a while and then the track—or so it seemed to some of us—was ripped up with indecent haste. A source of delight and usefulness had come to an end; and things haven't been quite the same since.